Applied Systems Thinking for
Health Systems Research

Applied Systems Thinking for Health Systems Research

A Methodological Handbook

Edited by Don de Savigny, Karl Blanchet
and Taghreed Adam

Open University Press

Open University Press
McGraw-Hill Education
8th Floor, 338 Euston Road
London
England
NW1 3BH

email: enquiries@openup.co.uk
world wide web: www.openup.co.uk

and Two Penn Plaza, New York, NY 10121-2289, USA

First published 2017

A catalogue record of this book is available from the British Library

ISBN-13: 978-0-335-26132-1
ISBN-10: 0-33-526132-9
eISBN: 978-0-335-26133-8

Library of Congress Cataloging-in-Publication Data
CIP data applied for

Typeset by Transforma Pvt. Ltd., Chennai, India

Fictitious names of companies, products, people, characters and/or data that may be used herein (in case studies or in examples) are not intended to represent any real individual, company, product or event.

Printed and bound by CPI Group (UK) Ltd, Croydon, CR0 4YY

Praise for this book

"For those working in the health sector, the relevance and value of systems thinking as a concept is evident. However, operationalization of this concept has been a challenge. With this new book, health researchers have a detailed guide for applying system thinking tools in day-to-day operations to identify and solve issues related to health policy and systems."

Ghaffar Abdul, *Executive Director of the Alliance for Health Policy and Systems Research, Switzerland*

"This keenly-awaited book offers a lucid and comprehensive discussion on how to research complex health systems. Health systems are facing a rapid change and increased complexity, with well-designed solutions often leading to unintended consequences. The book provides invaluable help in navigating this complexity and applying rigorous as well as pragmatic approaches to capturing dynamic interactions between system elements and causal loops. The authors operationalise systems thinking in an accessible style, presenting a menu of sophisticated methods and tools, and providing guidance on how to choose methods that fit the questions asked. This and the examples of practical applications of each method makes this book a key learning resource for health systems researchers, commissioners of research, programme implementers and equally, for any readers with interest in complex social systems."

Dina Balabanova, *PhD, Associate Professor, Health Systems/Policy, London School of Hygiene & Tropical Medicine, UK*

"Recognition of the need to enhance health systems around the world, instead of only focusing on diseases, has grown significantly over past decade or more. Consequently, production of health policy and systems related research also increased. However, going into an era of Sustainable Development Goals, studying and understanding how the national systems function from a systems perspective becomes even more necessary to find adequate remedies or policy solutions. This book will bring immense value to researchers and decision makers as it provides guidance not on "what to do" but rather "how to do" research from a systems thinking perspective. Systems thinking that emphasizes the context, complexity, interconnectedness, and adaptiveness of the health system resulting from feedback loops is gaining more traction within the health systems research field. Therefore, this book is timely to help equip individuals or researchers with practical tools that are expected

to empower the field and facilitate greater and more relevant evidence production for better policy making in health."

George Gotsadze, Director, Curatio International Foundation

"To ensure that health system research responds to policy priorities and to promote the use of evidence in policy decisions, researchers and health system managers should apply system thinking methods to address problems. This methodological handbook includes variety of essential tools and provides excellent guidance for health system researchers and managers when investigating and resolving complex health systems problems. Given the interconnectedness between health and other systems, the application of the tools included in this handbook is even more important when investigating health systems problems that are linked to the Sustainable Development Goals (SDGs)."

Fadi El-Jardali, Professor of Health Policy and Systems, American University of Beirut, Lebanon

"This book is the first to present practical options for applying systems thinking to understand complexity in health systems. The editors compile an essential collection of practical tools for understanding complex problems and framing research questions, as well as for determining and managing related solutions. Each tool is presented through an accessible summary of the method and the theory upon which it is based, as well as a real-world example. It will be a valuable resource for teaching and practice."

Ligia Paina, PhD, Assistant Professor, Bloomberg School of Public Health, Johns Hopkins University, USA

Contents

List of editors and contributors

Samuel D. Allen, PhD student, Worcester Polytechnic Institute, UK

Karl Blanchet, Associate Professor in Health Systems Research, London School of Hygiene and Tropical Medicine, UK

Don de Savigny, Professor, Health Systems and Policies Research, Swiss Tropical and Public Health Institute, University of Basel, Switzerland

Jenna M. Evans, PhD, Staff Scientist, Enhanced Program Evaluation Unit, Cancer Care Ontario; Assistant Professor, Institute of Health Policy, Management and Evaluation, University of Toronto, Canada

Glenda H. Eoyang, Executive Director, Human Systems Dynamics Institute

Lucy Gilson, Professor and Head of the Health Policy and Systems Division, School of Public Health and Family Medicine, University of Cape Town – and Professor of Health Policy and Systems, Department of Global Health and Development, London School of Hygiene & Tropical Medicine, UK

Hyunjung Kim, Associate Professor, California State University, USA

Kathy Kotiadis, Reader in Management Science, Christ Church Business School, Canterbury Christ Church University, UK

Annabelle Mark, Professor Emerita in Healthcare Organisation, Middlesex University, UK

Dintle S. Molosiwa, Post-doctoral Research Fellow at the University of Cape Town, School of Public Health and Family Medicine, Health Policy and Systems Division, South Africa – and Lecturer at Boitekanelo College, Department of Healthcare Service Management, Botswana

Daniel Cobos Muñoz, Health Systems Analyst, Health Systems and Policies Research Group, Swiss Tropical and Public Health Institute, University of Basel, Switzerland

Jill Olivier, Senior Lecturer and Research Coordinator at the University of Cape Town, South Africa, in the School of Public Health and Family Medicine, Health Policy and Systems Division

Erik Pruyt, Associate Professor in Model-Based Policy Analysis, Faculty of Technology, Policy and Management, Delft University of Technology, Delft, The Netherlands

Martin Reynolds, Senior Lecturer and Qualifications Lead for Postgraduate Systems Thinking in Practice, The Open University, UK

Vera Scott, Senior Researcher at the School of Public Health, University of the Western Cape, South Africa

Jessica Shearer, Senior Technical Officer, Health Systems Analytics at PATH, USA

Dave Snowden, Professor, Director Cynefin Centre, Bangor University, Wales, UK

Andrada Tomoaia-Cotisel, PhD Candidate, The London School of Hygiene and Tropical Medicine, UK; Policy Researcher, The RAND Corporation

Helen Wilding, Development Lead for 'Wellbeing for Life', Newcastle City Council, UK and Associate Lecturer, at The Open University, UK

Horst Christian Vollmar, Professor for Health Services Research, Institute of General Practice and Family Medicine, Hospital Jena, Friedrich-Schiller-University, Germany

1 Introduction: scope, intended audience, and how to use this handbook

Don de Savigny, Karl Blanchet and
Taghreed Adam

Foundations

Health systems are complex adaptive systems. They are thus characterized by extraordinary complexity in relationships among stakeholders and the processes they create. Systems phenomena of massive interdependencies, self-organizing and emergent behaviour, non-linearity and lagged feedback loops, path dependence and tipping points, make health system behaviour difficult and sometimes impossible to predict or manage. Conventional reductionist approaches to epidemiologic and implementation research are inadequate for tackling the problems health systems pose. It is increasingly recognized that health systems and policy research needs a special set of approaches, methods, and tools that derive from systems thinking perspectives (Adam and de Savigny, 2012; de Savigny and Adam, 2009), hence this handbook.

Health systems research is now finding its place at the heart of new developments in global health and has experienced marked expansion since 2010, the date of the first International Symposium on Health Systems Research. This health systems research boom is part of a new and still evolving discipline that has further revealed the complex adaptive nature of health systems.

The growing importance of health systems research was made possible by the first clear definition and conceptualization of the health system, including health system goals and values that were introduced by the World Health Organization in 2000 (WHO, 2000). This was further anchored by WHO in 2007 when they articulated more precisely a framework of six major sub-systems to provide practical boundaries of the health system (WHO, 2007). This in turn provided an opportunity for WHO and the Alliance for Health Systems and Policy research to introduce and promote systems thinking as a means to better understand health system behaviour (de Savigny and Adam, 2009).

Systems thinking and systems science are relatively familiar in other scientific disciplines, and to some extent, in disease-specific applications in the health sector, but the idea in 2009 to use systems thinking more broadly for health systems strengthening was new and has generated much interest. This interest has fuelled a demand for systems thinking methodological approaches and tools and the need to bring additional tools into the arsenal of health systems research (Carey et al., 2015; Swanson et al., 2012). One of the main questions asked by national health authorities, international donors, and research practitioners concerns how to conduct health systems research (HSR) from a systems thinking perspective (Peters, 2014).

From 1990 to 2009, there were only a handful of publications per year on systems thinking for health. Since then there has been an exponential rise in interest in applying systems thinking and systems science to complex adaptive health systems, with several hundred publications by 2017 (Carey et al., 2015; Chughtai and Blanchet, 2017). This now provides a sufficiently rich empirical experience from a number of real-world applications of systems thinking approaches and tools in health systems research to now make a methodological handbook possible. However, the full range of systems methodologies is yet to be engaged with by public health researchers (Carey et al., 2015).

Why are systems thinking tools important for health systems research?

The behaviour of complex health systems is determined by the actions of the actors and institutions involved and the power and politics that operate between them. Systems thinking places an emphasis on the interactions and connections among the different actors that make up a 'system'. A complex adaptive system can be described as a system that adapts and evolves through self-organization of a complex set of interconnected parts that operate for a defined purpose (Adam et al., 2009; Atun, 2012; Best and Holmes, 2010; de Savigny and Adam, 2009). Systems thinking emphasizes the importance of feedback loops and process contexts that make systems behaviour 'messy'. Research conducted at the intersection of health systems and health policy is by nature concerned with system functioning and policy change (rules of the system). A systems thinking approach helps such research move beyond the conventional linear approach or even the circular approach of research to policy to implementation to evaluation to further research, etc. Health systems research from this perspective is also at odds with linear models of the policy process. Many of the scholars who have critiqued the dominant approach to studying health policy and themselves utilized more non-linear policy process approaches (notably Gill Walt and Lucy Gilson) have become some of the core researchers within the health systems and policy analysis literature (Hanlin and Andersen, 2016). Health systems research from a systems thinking perspective is therefore more than research on health systems or health policy and can be differentiated from implementation research or operations research in that way (Gilson, 2012; Gilson et al., 2011).

Systems thinking approaches and tools covered in this handbook

The systems thinking research methods and approaches in this handbook can be loosely ordered and described under two overlapping themes: (1) tools for identifying and understanding problems and framing the research questions, and (2) tools for determining and managing solutions. The following is a brief guide to what is covered in the following methods chapters.

In Chapter 2, Jill Olivier, Vera Scott, Dintle Molosiwa and Lucy Gilson set the stage for this handbook with an essay on 'Systems approaches in health systems research: approaches for embedding research', where they address the emerging concept of embedding research in the health system itself. This *embedded systems approach* is more likely to lead to useable results and action by more appropriately identifying the problems and research questions. In so doing, such research enhances accountability for both the evidence produced and the decision-making that should follow, since the researchers and policy-makers/managers work together to hold each other accountable (WHO, 2012). This chapter therefore addresses a wide spectrum of themes important to the systems thinking approach to health systems research with its participatory focus on identifying and understanding problems and stakeholder relationships, its highly reflective nature, and its emphasis on negotiating solutions and problem-solving.

Embedded systems approaches should be of particular interest to health systems and policy researchers who find themselves working in health systems as 'insider-researchers', including student researchers, jointly appointed staff, and those acting in research partnerships and joint work programmes, or as part of NGO/donor agency-funded staff placed inside the system. This chapter richly describes the many advantages of the embedded approach, along with the challenges and specific ethical considerations. It concludes, as all chapters in this book do, with a worked, real-world case example. The chapter thoroughly documents an experience from Africa describing the application of an embedded approach to district innovation and action learning for health systems strengthening.

In Chapter 3, Martin Reynolds and Helen Wilding introduce 'Boundary critique: an approach for framing methodological design'. Whenever we talk about health systems and systems research, there is the question of boundaries. Health systems are nested within wider socio-political-economic-legal systems, and health systems nest within them further sub-systems (most notably service-delivery systems, human resource systems, information systems, etc.), which in turn nest further sub-systems (hospital care systems, etc.). One of the first steps in health systems research is reflecting and deciding on the boundaries of the system in question, both explicitly and implicitly. Boundary technique is not so much a methodology as it is a necessary reflective systems approach to methodological design. It does so in part by facilitating triangulated reflection on judgements of fact, value judgements and boundary judgements using various approaches and tools such as critical systems heuristics. Chapter 3 provides the history and theory of boundary technique and provides a practical example of its application. The case provided concerns partnership arrangements among the many partners operating in the UK National Health Service reforms where there was a broad range of multi-finality perspectives of the partnership. Boundary technique is one of the family of soft systems methodologies further described in Chapter 4.

Chapter 4, by Kathy Kotiadis, is titled 'Soft systems methodology: an approach for stakeholder and researcher reflection on a problem'. There is a great deal of interest in how the application of systems concepts and approaches might aid public health. A recent analysis suggests that 'soft systems modelling techniques are likely to be the most useful addition to public health, and align well with current debate around knowledge transfer and policy' (Carey et al., 2015). As exemplified in Chapter 3, many stakeholders (partners) in a system may reach very different opinions about the cause and resolution of a system problem. This is because most actors in a complex adaptive system are focused on their part of the system and have insufficient understanding of the whole system. Soft systems methodology is an organized learning system that facilitates a shared view of the situation and the development of agreement towards desirable action. It is an approach rather than a tool *per se* that offers a process of enquiry when there is ambiguity or disagreement about the problem. It is a package of simple models of practice that can be facilitated by insider-researchers (Chapter 2) or external analysts. The first part of the chapter provides a thorough review of the origins, history, and theory behind the approach, while the second part describes an actual application. The application illustrated concerns a multidisciplinary cancer team of surgeons in a hospital who wanted to evaluate how their team functioned to maximize benefit for patient outcomes.

In Chapter 5, Annabelle Mark and Dave Snowden introduce 'Cynefin: a tool for situating the problem in a sense-making framework'. Cynefin (pronounced Kun-ev'in) is a Welsh word denoting the 'place of your multiple belongings' (cultural, geographic, religious, tribal) that affects the 'sense of being part of a flow of meaning over time . . . that profoundly impacts on how we think and act'. The authors are the original architects of this decision framework tool that helps to make sense of the system and its problems sufficiently to know enough to act. The framework in part helps us describe a problem based on self-signified micro-narratives of lived experience of stakeholders in the system (the sense-making mechanism), and then situate the problem within boundaries that reflect distinct categories of systems as being either *ordered* (simple or complicated) or *unordered* (complex or chaotic) and sometimes *disordered* when at the nexus of these domains. These framework boundaries then provide a guide to the different types of action, analysis, and practice. This chapter richly describes Cynefin and its methods, including links to software and video support. It concludes with a worked example illustrating how the Public Health Agency and the Health and Social Care Board in Northern Ireland used it in a partnership approach to improve patient and client experience.

In Chapter 6, Andrada Tomoaia-Cotisel, Hyunjung Kim, Samuel Allen and Karl Blanchet address a classic systems thinking tool, that of 'Causal loop diagrams: a tool for visualizing emergent system behaviour'. The causal loop diagram (CLD) is a visual representation of causal linkages among elements of a system in order to understand a specific problem. CLD has the advantage of creating a mental model of a health system that can be shared and discussed between stakeholders. It also takes into account the dynamics of health systems by illustrating feedback processes, non-linear relationships between elements of a system, and time delays between cause and effect. Visual representation in CLDs helps stakeholders see the nature of the dynamic complexities in the system and, in so doing, better understand their policy implications. The methodology involves a step-by-step process that consists in identifying a causal relationship between two elements in the system that are represented by two nodes (i.e. variables), and an arrow connecting those two nodes with a polarity. Two examples are provided from two studies, one carried out in New Zealand and the other in Uganda.

In Chapter 7, Karl Blanchet and Jessica Shearer tackle 'Network analysis: a tool for understanding social network behaviour of a system'. Social network analysis (SNA) is about understanding the influence of interpersonal relationships, and the larger structure they create, on decision-making and a range of other individual- and system-level behaviours. Individuals are connected to other individuals, forming a social network. Any research question related to networks is situated on at least one of these levels: individuals ('nodes'), relationships ('ties'), and the resulting network/system. A basic approach to SNA captures all three levels of the network during the following three stages: (1) describing the set of actors in the network, (2) characterizing the relationships between actors, and (3) analysing the structure of the systems. More complex methodologies such as exponential random graph models and separable temporal exponential random graph models have been applied to describe the time-sensitive dynamics of health systems. The application of SNA is explained through two case studies, one conducted in Ghana and the other in Burkina Faso.

Chapter 8, by Glenda Eoyang, focuses on 'Human systems dynamics: a tool for understanding self-organizing behaviour of actors in the system'. Health systems are human systems, and any self-organizing behaviour within the system is driven by internal tensions of the actors, whether at the level of the individual, team, institution or community. These form the dynamics of the system that while unpredictable and difficult to control, can still be observed, anticipated (with systems thinking), and influenced. This chapter provides six tools derived from the theory of non-linear dynamics that are relevant to health systems research. They are: the Container-Difference-Exchange (CDE) model for self-organizing in human system, Pattern Logic, Adaptive Action, Same & Different, Landscape Diagram, and Designing Exchanges. A case example of the use of the human systems dynamics tools in action is provided in a case of adaptive action learning to respond to system challenges of medical professional teams, their patients, and organization.

In Chapter 9, Daniel Cobos Muñoz and Don de Savigny introduce 'Process mapping: a tool for visualizing system processes from end-to-end'. Systems thinking requires us to understand the whole, not just the parts, and to understand the relationships among the parts, including the stakeholders (see also Chapter 7). Process mapping is a tool that helps illuminate all of this in an easily understandable visualization. Actors in a system engaged with complex processes may understand their own work well, but will have a less clear picture of how it fits in the end-to-end flow of the process across all the stakeholders concerned. Process mapping helps to understand stakeholder relationships, identify bottlenecks, inefficiencies and design flaws that limit system performance. Process maps also help to think through the system re-engineering need to integrate new interventions in the system and to dynamically model system performance and manage change over time. Process mapping is an effective tool for communicating, analysing, sense-making, and managing. This chapter illustrates all of these functions and provides a 'how to' strategy for implementation. It concludes with a real-world application where process mapping has been used to help stakeholders understand and re-engineer the complex health information sub-system of civil registration and vital statistics for births and deaths in a low-income country working across three national ministries.

In Chapter 10, Erik Pruyt addresses 'Systems dynamics: a tool for modelling and testing solutions'. Systems dynamics is a methodology for modelling and simulating the dynamics of complex systems to better understand the whole system, gain insights into system structure and system behaviour, discover potential future system dynamics, and identify high-leverage policies and systemic interventions. As such, it can be applied to many health, healthcare, and health system issues. Systems dynamics also frequently makes use of causal loop diagrams (Chapter 6) and stock-flow diagrams. This chapter provides an introduction to systems dynamics and illustrates the use of systems dynamics initially with the relatively simple case of a disease outbreak. It then provides a more complex example of modelling in health systems. In this case, how the health system responds to a particular health service (peripheral vascular disease).

Chapter 11, by Horst Christian Vollmar, is titled 'Scenario technique: a tool for simulating and reflecting on alternative solutions'. The scenario technique has had a resurgence of application since 2000 with over 40 journal articles using the technique in the context of strategic planning for healthcare systems, although only one of these in a low-income country. Scenarios are causally coherent, internally consistent, and hypothetical, descriptions of alternative futures that reflect different perspectives on past, present and planned developments, which can serve as a basis for action. Scenarios are not predictions. Horst describes three main approaches to developing and using scenarios, along with the limitations of the technique. Its application is illustrated by two cases: one from a high-income country (health care for dementia in Germany) and one from a low-income country (community health development in Vietnam).

Chapter 12, the final tool chapter, is by Jenna Evans, who addresses 'Outcome mapping: a tool for planning, monitoring, and evaluating complex interventions in systems'. Most funding partners of health systems interventions and health systems research require a planning, monitoring, and evaluation plan based on classic logframe tools, which see project implementation and development in a linear, one-way, cause-and-effect process (inputs, outputs, outcomes, impacts with specified assumptions and verifiable indicators). Unfortunately, health systems are not linear. Outcome mapping is a systems thinking alternative to planning, monitoring, and evaluation that recognizes that that there are multiple non-linear causes and pathways to change, and that the funded intervention merely contributes to this, but is not necessarily responsible for it. Outcome mapping therefore takes a more inclusive perspective on the stakeholder engagement (boundary partners) and their contribution and role in consensus-building for a shared vision of how collective activities lead to outcomes. Finally, outcome mapping is concerned more broadly with changes in behaviours and relationships of stakeholders, and not just narrowly defined 'hard' deliverables for the primary beneficiaries. For these three reasons, outcome mapping clearly qualifies as a systems thinking tool. In this chapter, the foundations of outcome mapping are reviewed and the methodology clearly laid out. Two case examples are provided to illustrate its use. The first concerns the application of outcome mapping in a programme in South Africa designed to improve delivery of primary health care at district level, and the second concerns its application in Canada to achieve stakeholder alignment around a shared vision to integrate health service plans in local Health Integration Networks involving local healthcare providers and community members.

In the final chapter, 'Conclusion: mapping methods to research challenges', the editors tackle the challenge of mapping these systems thinking methodologies to health system research issues. We provide a graphical matrix as a guide to these methods and approaches that situates them with regard to where they fit in the sequence of research steps that include problem identification, participation and reflection, setting boundaries, understanding problems, visualization, understanding dynamics, understanding stakeholders, sense-making, negotiating solutions, testing solutions, problem-solving, and framing outcomes.

Intended audience

This handbook is aimed at providing practical guidance to students, researchers, and professionals who want to apply emerging systems thinking approaches and methods in health systems research. It is intended to be a practical handbook and learning tool, covering well-tested and practical aspects of systems thinking methods, tools, and techniques useful in health systems research. It has been designed to be a comprehensive 'how to guide' for students, academics, research users, and other health professionals. Its purpose is to prepare researchers and system managers to apply these principles and practices immediately to their own complex health system challenges.

How to use this handbook

In this book, we present a series of empirically underpinned methods chapters describing some of the main innovative systems thinking approaches, methods and tools being used in health systems research today. Each chapter is written by an individual or a team of authors identified by the editors and selected for their level of applied expertise. Each chapter is in two parts. The first starts with a historical overview of the method, its theoretical underpinnings, and a clear description of the method. The second part provides a practical worked real-world example of an application of the method to illustrate how it can be implemented to tackle a relevant health system research challenge.

References

Adam, T. and de Savigny, D. (2012) Systems thinking for strengthening health systems in LMICs: need for a paradigm shift, *Health Policy and Planning*, 27 (suppl. 4): iv1–iv3.

Adam, T., Mookherji, S., Campbell, S., Reid, G., Gilson, L. and de Savigny, D. (2009) Systems thinking for health systems: challenges and opportunities in real-world settings, in D. de Savigny and T. Adam (eds.) *Systems Thinking for Health Systems Strengthening* (pp. 71–86). Geneva: Alliance for Health Policy and Systems Research/World Health Organization.

Atun, R. (2012) Health systems, systems thinking and innovation, *Health Policy and Planning*, 27 (suppl. 4): iv4–iv8.

Best, A. and Holmes, B. (2010) Systems thinking, knowledge and action: towards better models and methods, *Evidence and Policy*, 6 (2): 145–59.

Carey, G., Malbon, E., Carey, N., Joyce, A., Crammond, B. and Carey, A. (2015) Systems science and systems thinking for public health: a systematic review of the field, *BMJ Open*, 5 (12): e009002 [doi: 10.1136/bmjopen-2015-009002].

Chughtai, S. and Blanchet, K. (2017) Systems thinking in public health: a bibliographic contribution to a meta-narrative review, *Health Policy and Planning* [doi: 10.1093/heapol/czw159].

de Savigny, D. and Adam, T. (eds.) (2009) *Systems Thinking for Health Systems Strengthening.* Geneva: Alliance for Health Policy and Systems Research/World Health Organization.

Gilson, L. (ed.) (2012) *Health Policy and Systems Research: A methodology reader.* Geneva: Alliance for Health Policy and Systems Research/World Health Organization.

Gilson, L., Hanson, K., Sheikh, K., Agyepong, I.A., Ssengooba, F. and Bennett, S. (2011) Building the field of health policy and systems research: social science matters, *PLoS Medicine*, 8 (8): e1001079.

Hanlin, R. and Andersen, M.H. (2016) *Health Systems Strengthening: Rethinking the role of innovation* [retrieved from: http://aauforlag.dk/Shop/organisation-og-ledelse-(1)/health-systems-strengthening-rethinking-the-r.aspx].

Peters, D.H. (2014) The application of systems thinking in health: why use systems thinking?, *Health Research Policy and Systems*, 12: 51 [doi: 10.1186/1478-4505-12-51].

Swanson, R.C., Cattaneo, A., Bradley, E., Chunharas, S., Atun, R., Abbas, K.M. et al. (2012) Rethinking health systems strengthening: key systems thinking tools and strategies for transformational change, *Health Policy and Planning*, 27 (suppl. 4): iv54–iv61.

World Health Organization (WHO) (2000) *World Health Report 2000: Health systems: Improving performance.* Geneva: WHO.

World Health Organization (WHO) (2007) *Everybody's Business: Strengthening health systems to improve health outcomes: WHO's framework for action.* Geneva: WHO.

World Health Organization (WHO) (2012) *Changing Mindsets: Strategy on health policy and systems research.* Geneva: WHO.

2 Systems approaches in health systems research: approaches for embedding research

Jill Olivier, Vera Scott, Dintle Molosiwa and Lucy Gilson

Introduction: an embedded approach to HPSR

The World Health Organization (WHO) strategy on health policy and systems research (HPSR) calls for an 'embedded' approach to work in this field, seeing HPSR as a core function of every system (WHO, 2012; see also Ghaffar et al., 2017; Hoffman et al., 2012).

An embedded approach becomes relevant as a result of two main characteristics of HPSR. First, HPSR is said to be problem-driven, focusing primarily on solving complex real-world problems. HPSR is therefore very much an applied field, whose problems shape the questions asked and then the methods that get applied. Research that is properly embedded in the health system is more likely to lead to actionable and useable results by appropriately identifying critical problems and questions. Indeed, in HPSR, the 'substantive relevance' of the research is a core consideration for researchers – that is, whether the research is 'worth doing', with worthiness mainly judged in terms of its value for system development and its relevance to health systems actors (decision-makers at all levels of the system, including at community level; see Dunston et al., 2009; Ghaffar et al., 2017; Gilson, 2012). Second, HPSR is centrally concerned with 'people', their relationships and actions (see Gilson, 2012; WHO, 2012). Health system development and HPSR are understood to be socially embedded, with knowledge production and use shaped by people (see Green and Bennett, 2007). This focus on people suggests that: 'HPS researchers must work in direct engagement with the practitioners and practice central to the inquiry, acknowledging their tacit knowledge and drawing it into generating new insights into health system functioning' (Lehmann and Gilson, 2015: 957).

Despite the growing importance placed on 'embeddedness' in HPSR (see Ghaffar et al., 2017), this is still a largely unexplored area for the field. In the existing HPSR literature and discourse, embedded research is mainly described as an issue of *organizational embeddedness* leading to improved research translation and utilization by health systems decision-makers. For example, a background framing document on embedded research commissioned by the Alliance for Health Policy and Systems Research (AHPSR) focused on what they called 'institutional embeddedness' (Koon et al., 2012). The focus here appears to be on national-level strategies for strengthening evidence-based policy and decision-making, mainly by ensuring that research results are strategically placed and communicated outwards to higher-level health systems actors – usually by ensuring a stronger relationship between research institutions and national Ministries of Health (see Koon et al., 2012; WHO, 2012).

This institutional and macro-level communication and utilization of evidence is certainly important. However, there is little guidance on what embeddedness means on a methodological level – for researchers seeking to practise effective embedded HPSR at all levels of the health system. Yet for many types of HPS researchers, an embedded approach is immediately relevant and sometimes a matter of daily concern. As Table 2.1 illustrates, embedded researchers might be: practitioners or health workers inside the system conducting HPSR (often called 'insider-researchers' in the literature);[1] students working in the system and conducting research relating to their work and also for their student project; staff with joint appointments in government positions and in academia; research partnerships and joint programmes of work (such as learning sites); longer-term institutional relationships (not always project specific); longer-term HPSR project researchers from outside institutions who immerse themselves within the system for a finite period of time (such as ethnographers and action researchers); or NGO/donor agency-funded staff placed inside the system (usually for a particular purpose or programme of work).

These examples highlight common HPSR scenarios that are several steps back from macro-level institutional arrangements and evidence translation. They represent *the regular practice of embedded research*, or an embedded research approach, which thousands of HPS researchers around the world are engaging in on a daily basis – and they are the focus of this chapter. There is little guidance for HPS researchers setting out to *practise* embedded HPSR, seeking ideas about appropriate modes of practice or advice on the mechanics of embedded research in HPSR.

In this chapter, we draw on existing literature and our experience of HPSR to clarify what we judge to be the key features of an embedded approach in HPSR, and then present one case example of this approach from our research experience in South Africa. We stress that embedded HPSR is best understood as a *research approach*, not a specific method (such as ethnography) or tool (such as causal loop diagramming). It is an approach or research 'style', which is about practising HPSR with a particular ethos, accepting the socially constructed nature of health systems and policies, and locating the research within a particular relationship between practitioners and researchers. Within an embedded approach, a range of methods and tools may be utilized. So while it is easier to describe an embedded ethnography or to identify an embedded flexible design (including any of those described in this book), a fixed-design[2] randomized control trial *could* form part of an embedded approach, if it is properly negotiated and framed within the type of research relationship outlined below.

An embedded approach contains many elements of action research and ethnography, but does not encompass either. For both of these, there is already extensive literature on the method and challenges inherent in these approaches for HPSR, and we therefore only summarize these here (see Loewenson et al., 2014). Given that an embedded approach encompasses other methods and tools,

Table 2.1 Types of embedded researchers commonly found in HPSR

Type	Description	Examples and exemplar outputs
Insider-researchers	Practitioners (health workers) working in the system, conducting health systems research	Internal policy documents and reports (not public)
Insider *student research*	Health system workers also registered for study with an academic institution – usually conducting research on issues related to their work	For example, as seen in published theses, and some HPSR PhD student publications (e.g. Koduah et al., 2015)
Jointly appointed staff	Jointly appointed staff working in the health system and in academia, as part of an institutional arrangement (or secondment)	For example, the work of Irene Agyepong, who has held joint appointments within the Ghanaian Health Service and the University of Ghana (and other institutions; see Agyepong et al., 2012)
Research partnerships and joint programmes of work	Longer-term partnerships and arrangements either set up around a specific programme of work, or a series of different smaller projects	Partnerships supported by PAHO and AHPSR though the iPIER (Improving Programme Implementation through Embedded Research) initiative implemented in the Americas Learning sites set up through the RESYST consortium in South Africa, Kenya and Tanzania (see RESYST, 2016) The CHESAI collaborative in the Western Cape of South Africa, building joint learning between academic and local health system actors (see CHESAI, 2016)

(Continued)

Table 2.1 continued

Type	Description	Examples and exemplar outputs
HPSR project researchers immersed in the system	Longer-term HPS research projects where researchers from outside institutions immerse themselves within the system for a finite period of time	Ethnographers such as Asha George, whose work 'By papers and pens' (2009) describes work over five years in India Action researchers with a high level of participation in the system (e.g. Khresheh and Barclay, 2007) Ongoing networks engaged in action research such as EQUINET (see Loewenson et al., 2014)
NGO/donor agency-funded research staff placed inside the system	Researchers or managers seconded to or embedded within a system for a particular purpose or programme of work	In South Africa, for example, PEPFAR-USAID has funded the insertion of staff into the Ministry of Health to work on HIV and PHC integration Also in India, where over 40 researchers were embedded in the National Ministry of Health (and then later removed)

the ethical and rigour-related research guidance we give here relates to the over-arching approach, and researchers will still need to address the specific method-ological and ethical considerations for any other methods or tools that they draw into an embedded approach. For example, if a case study method is utilized within an embedded approach, the challenges specific to case study methodology still need to be observed (see Gilson, 2012).

Finally, it is necessary to stress that we do not consider embedded research to be the sole territory of social scientists working in HPSR. To the contrary, it is more frequently those with a health sciences or managerial background, with little previous exposure to this approach, that tend to find themselves wrestling with the challenges of an embedded positionality within their health system. HPS researchers with a social science background may more easily navigate the tensions inherent in an embedded approach, and have usually received training and the tools that help this navigation, while others tend to gain these competen-cies through experience. In keeping with the understanding that HPSR is an inter-disciplinary field, we address this chapter to 'researchers' of all backgrounds and epistemologies, with the expectation that everyone in HPSR is likely to find them-selves in a situation of embeddedness at some stage of their career. This is an introductory chapter – and we provide references for those seeking a more in-depth engagement with the issues raised here. We also encourage further pub-lication of real-world embedded HPSR experiences, so that the practice becomes more fully developed within HPSR (see Caffrey et al., 2016).

What is (and what is not) an embedded approach to HPSR

'Embeddedness' is a term used widely across a number of different disciplines, and there is no single definition. Dacin et al. (1999) note that the term has taken on multiple meanings in the same way that 'culture' and 'trust' have. A general search of 'embedded research' reveals several different areas of investigation that have relevance to HPSR.

In the most basic sense, 'embedded' is used to describe how something is placed within something else. For example, in HPSR, the term is often used to describe how software or values are embedded within health systems, such as when cer-tain values drive or are embedded in the idea of universal health coverage (UHC; see Gilson, 2012; WHO, 2012). Blanchet and James (2012) consider how health systems are seen as combinations of systems embedded within each other (public/private, local/global, social/organizational), and because of this embeddedness an event at one level can have an impact at another level.

More pragmatically, an embedded approach is commonly used to describe a 'nested' method, such as an embedded case study approach (a type of case study where one case is nested within another), an embedded mixed-method approach (where one methodological component is integrated into another), or an embed-ded evaluation (where evaluation is built into programming or teaching so that it accompanies implementation rather than added on at a later stage).

Within the social sciences, theories of embeddedness are common. 'Social embeddedness' is often understood as representing an organization's or individual's connection, relationship and/or position within a social network (see Nee and Ingram, 1998). This concept is applied in a number of different ways (summarized by Koon et al., 2012), such as in the analysis of social networks or the examination of informal collaboration. Within this body of work, there are two dominant conceptualizations: 'embedded theory' in social economics (see Talmud, 2013), and the positioning of organizations within networks – where it is usually shown that more embedded organizations tend to be more trusted and have more influence to effect change (see Cook et al., 1983; Koon et al., 2012). As noted earlier, this latter focus seems to have been influential for the WHO-AHPSR Background document, which focuses mainly on the embeddedness of *research institutions* within systems, and their resulting influence on high-level decision-making (Koon et al., 2012).

All of this is relevant, but does not entirely encompass the full characteristics of socially embedded HPSR – or the considerations required for an embedded research approach carried out by researchers who are also health systems actors. Such researchers are characterized by their 'situatedness' within a health system, the influence of their interpretation of the system around them (on their research), and the potential change they can effect (even just by asking questions). With such researcher-actors in mind, we understand an embedded approach to be one in which *researchers negotiate and conduct research from within the health system that is the object of their study (positioned as insiders), usually with the intention that their research will lead to positive health systems change.*

We now identify characteristics of an embedded research approach, and address key challenges and researcher competencies. Some HPS researchers may foreground embeddedness more than others – or particular projects may be more easily embedded than others. We therefore present these as a cluster of defining characteristics – and an HPSR study might well be considered 'embedded' even if not all the characteristics are equally present.

The researcher behaves as an 'insider-researcher' within their health system

The WHO HPSR strategy observes that embedded HPSR is different from other forms of public health research, 'which are generally more distant from the object of study in order to preserve scientific integrity' (2012: 14). An embedded approach, in contrast, specifically requires the researcher to be close to the object of study, working '. . . in direct interface with the constituency they serve; not remote and distant from the "object" of their study' (Lehmann and Gilson, 2015: 958). If you appreciate semantics, this would mean the embedded HPS researcher is conducting research 'in' the system, not 'on' the system – or describe it as 'their' system, not 'the' system.

This closeness is both physical and conceptual. Physical closeness is an important (and often ignored) factor, enabling the researcher more easily to observe and interact with the routine and real functioning of 'their' health system. At an institutional level, this is why Koon et al. (2012) advocate for research-producing institutions to be situated in close proximity to decision-makers.

However, the closeness to the research object is also conceptual – in this case meaning that an embedded approach requires the researcher to be (behave as, or

Box 2.1:   Idealized criteria for practitioner research

Reed and Procter (1995) identified 'idealized' criteria for practitioner research in health care as:

- a social process undertaken with colleagues
- educative for all participants in the projects
- imbued with an integral development dimension
- focused upon aspects of practice in which the researcher has some control and can initiate change
- able to identify and explore socio-political and historical factors affecting practice
- able to open up value issues for critical enquiry and discussion
- designed to give a say to all participants
- able to exercise the professional imagination and enhance the capacity of participants to interpret everyday action in the work setting
- able to integrate personal and professional learning
- likely to yield insights that can be conveyed in a form that make them worthy of interest to a wider audience.

be viewed as) an 'insider' within their health system. Outside of HPSR, especially in the area of educational studies, a substantial body of work addresses 'insider research' (also called 'practitioner research' or 'work-based research'; see Box 2.1). Insider research is usually defined as research conducted within an organization where the researcher is also employed by that organization. We would argue that in HPSR, health systems are understood to encompass a broad range of actors. That is, actors and decision-makers in health systems are not only policy-makers or managers in Ministries of Health, but are also other cadres of health workers, non-state and civil society actors, patients, and 'researchers'. All of these might be considered 'insiders' to a health system.

Insider-researchers in HPSR often wear multiple hats simultaneously (as researchers, health workers, decision-makers, and patients), and complex secondment and joint appointments make even employment or institutional affiliation blurry at times. This can be seen in the way that HPS authors commonly claim different institutional affiliations in different publications – not because they have moved between institutions, but because affiliation often requires negotiation as part of the research process and agreement. We therefore argue that researchers utilizing an embedded approach in HPSR would *behave as, or be reflexive about, their role as insiders to the health system* – whether they are formally employed within the system or not.

The benefits of conducting research from within the health system are numerous. These include: having more in-depth (insider) knowledge of the system and context in which the research is being conducted; a greater likelihood of identifying substantively relevant problems and questions; better access to people and information; less likelihood of being blocked by gatekeepers within the system; a greater chance to observe the routine functioning of the system; a greater likelihood of seeing and having access to tacit knowledge within the system; more opportunities to engage with difficult findings in safe spaces within the system (that is, negative or difficult findings are more likely to be discussed internally than if exposed by an outsider in a confrontational manner); greater opportunities to feed research findings more rapidly back into the system; and in some cases the embedded insider-researcher is in a better position to make the changes within the system that the research findings suggest are appropriate (see Costley et al., 2010; Gilson, 2012).

There are also challenges that emerge alongside these benefits. The most obvious is that insider-researchers might struggle to maintain objectivity, weakening research results and authenticity. For example, insider-researchers might feel compelled to report more positively on a finding than a third-party researcher would, or be pressured by the politics always present within health systems to report in a particular way. This is a concern that is widely discussed in relation to 'insider evaluations'.[3] Health systems are developed and maintained through power, and embedded researchers can sometimes be caught in such power dynamics (from which a third-party outsider researcher is usually more protected). Embedded researchers might be directly in a position of power, and there are obvious challenges involved in research situations such as when a manager interviews a subordinate colleague who is 'compelled' to participate and likely to adjust their responses according to what they think the manager wants to hear. Consider also the challenges inherent in researching a health systems intervention that the researcher was responsible for developing and implementing. (Such challenges are addressed through reflexive practice, addressed below.)

While being an insider might enable the researcher to access privileged knowledge and be aware of the tacit (unspoken) knowledge within the system, at the same time, being an insider, immersed in the system and context, might blind the researcher to certain norms and key issues. It is sometimes difficult for insiders to see 'the obvious' or 'the normal' from the inside. Costley et al. note: '. . . as an insider who is immersed in work, it is possible to fail to see the obvious and you need external feedback on what you are doing' (2010: 4).

Another challenge that is commonly raised is that the research boundaries are often blurred. For example, many insider-researchers, especially students, struggle to differentiate between 'observation' and 'experiential knowledge'. They are unsure whether a piece of information was a formal observation, or something they 'just know'. They might agonize over whether a piece of information was given to them as a researcher or as a manager, and what rules of confidentiality or permission might govern a particular conversation. In the real world of research, it is not always possible to pause and clearly delineate such boundaries, or clearly articulate to a colleague which hat you are wearing at a particular time (see Coghlan, 2007). Insider-researchers also sometimes find it impossible not to act on what they find during the research process. For example, if they find something failing or abuse being conducted, they are unlikely to be in a position to wait until the research process has formally ended before acting on it. For instance, what they learn would inform their everyday decision-making. This may result in ethical dilemmas (to act or not), and may also affect the research setting itself – changing the system while the research is underway.

We noted earlier that people and relationships are important in HPSR. However, these relationships can cause particular challenges in embedded research. In traditional research, researchers can more easily define the boundaries of relationships, and are trained to detach themselves from the research context. Insider-researchers are often unable to do this, as they are part of their system, have vested interests in it, and are usually in relationship and emotionally engaged with their 'study subjects'. Costley and Gibbs point out that these relationships raise specific ethical concerns, such as 'how should the researcher behave when the findings of the research might affect or even injure those to whom the research has a special professional, functional or emotional bond?' (2006: 89).

None of these challenges are new or specific to embedded research or HPSR. However, they are particularly apparent in embedded HPSR practice. Many solutions have been developed to manage these tensions, which have been usefully drawn from other areas. One of the main competencies for embedded HPS researchers is the adaption of a *reflexive approach*. This is often raised in research training; however, it is a complex competency that is developed slowly over time and through experience. In embedded HPSR, this might include: the researcher learning to be more transparent about their own positionality and subjectivity in the research process (and seeing this as an ongoing process in the research, rather than a quick reflection in an ethics application); being consistently reflective about the potential impact the research might have on the system; improved reflexive research and writing approaches which clearly articulate, for example, what roles and influences the researcher has in the system; 'checking' approaches such as drawing on an outsider perspective to check that the insider-researcher's positionality has not blinded them to important norms; and by becoming familiar with reflective practice tools and methods that can be applied repeatedly in the field.[4]

Costley et al. (2010) point out that such competencies require self-belief and confidence, as the insider-researcher is often required to justify their research and results to critical audiences in both academia and the workplace (in this case, the health system). It is necessary to note that some reflexive practice approaches (especially in the social sciences) result in a deeply self-critical discourse and approach, which does not always communicate across to those in the workplace or practice environment as confidence or self-belief. It is important that embedded researchers are trained to become confident in the value of their embedded approach, while remaining transparent about the challenges inherent in this approach. Such confidence is built by the researcher being clear in their understanding of the underpinning reasons for conducting embedded research in HPSR (see above), and by drawing on the wealth of methodological resources outlined in this chapter and the rest of this book.

The greatest challenge in insider-research is the issue of subjectivity versus objectivity – and the resulting perceived rigour of the resulting research. Again, this is a vast area of discussion within action research and ethnographic circles – and we cannot do justice to it here. Embedded researchers need not feel that they are alone in wrestling with these complexities. There are many articles addressing techniques for 'managing subjectivity' in research settings where pure objectivity is simply not possible, viable or desirable – as is often the case in HPSR (see Gilson, 2012; Green and Bennett, 2007). For example, Heshusius (1994) speaks of how to 'free oneself from objectivity' and discusses what managing subjectivity and a turn to a participatory mode of consciousness and research requires. In action research circles, there is also great debate on subjectivity versus objectivity and issues of rigour (see Loewenson et al., 2014). Again, we are not suggesting that all HPSR becomes participatory or ethnographic in nature. However, we highlight that any embedded HPS researcher (even those utilizing quantitative methods) will have to face up these issues – and will need to be able to argue the extent and limitation of findings emerging out of an embedded-style research approach.

Continuous negotiation, co-creation, collaboration, and trust-building between stakeholders

The identification, framing, and planning for research that addresses complex real-world problems from within health systems requires high levels of continuous negotiation between different system actors. An embedded approach in HPSR usually requires *high levels of negotiation* (for researchers seeking to become embedded, and those that are already embedded as a result of their employment). However, such skills are not often taught to researchers, and the extensive negotiation that usually precedes research and continues in the background through the research process is rarely reported on. We argue that in embedded HPSR, it is necessary to foreground this 'behind the scenes' negotiation, and consider this as a key element of the research process. Negotiation in embedded HPSR is usually something that has to happen before, during, and after the research process in a continuous cycle – where the researcher sets out to develop and maintain a fair, successful, and enduring research partnership. We highlight key phases and areas where negotiation is usually required.

Negotiation requires the proper identification of research partners. Embedded HPS research relies on the proper identification of research partners with whom to negotiate, something that is particularly challenging in health systems where there are a complex array of relevant actors, beyond high-level policy managers or the standard gatekeepers to insider-research in organizations (see Gilson et al., 2014; Lehmann and Gilson, 2015). Tools such as stakeholder analysis and actor mapping are useful to assist embedded researchers to map and understand actors better. Action research also has useful tools to turn research 'subjects' into active research participants (see Loewenson et al., 2014). In embedded HPSR, it is usually important that all relevant research partners (stakeholders) are identified and negotiated with early on in the research process. It is rare that only one level or type of actor from one institution would be identified as a research partner, and more likely that varied partners will be identified. It is also likely that the actor mapping and negotiation process will need to be repeated, as actors change frequently within health systems contexts.

Early negotiation of purpose/objectives between research partners: Embedded HPSR requires intense negotiation of the research objectives. Lehmann and Gilson identify a 'double hurdle of scholarly quality and relevance' (2015: 958) in HPSR, which highlights that in some situations, (academic) researchers and practitioners in the health system may have different objectives, and may assess relevance differently. Negotiation is required so that research partners can learn about the values, priorities, and practices of one another, and so that balance can be achieved and common ground found between different agendas (see WHO, 2012).

Early negotiation in problem identification and framing: In embedded research, substantial time is ideally allocated to early negotiation in which the issue, problem or question of focus is jointly identified and the research framed. Such negotiation underpins the development of substantively relevant research that draws on existing research capacities. There are, of course, many examples of real-world HPSR where the research question has been pre-set (e.g. by a funding agency or by the researcher). In such cases, it would depend on whether the research displayed other characteristics (as described here), as to whether it could still be considered to be an embedded approach.

Negotiation in setting up the parameters of the research partnership: In embedded research, it is usually necessary to clarify the terms of the research and the research partnership early and transparently. This is standard practice in all research, but in embedded research, it is necessary to go beyond the basic requirements of ethical committees and legal contracts. Instead, researchers need to openly and continuously (re)negotiate issues such as ownership and utilization of data and results, boundaries, and the nature of the research partnership. It is necessary to establish common purpose and mutual benefits; to express and manage expectations; to raise and anticipate any potential risks or unanticipated consequences (see below); and to clarify contributions of funding as well as in-kind costs such as information sharing and staff time. It is important that this negotiation is documented – both in standard terms of reference, contracts, and agreements, but also through internal sharing of documentation of early meetings and discussions (as these parameters will change during the research and this negotiation will often need to be revisited).

Negotiation needs to be consistently and flexibly reiterated: Because the process of inquiry in embedded research is often emergent, it requires greater flexibility and responsiveness among the engaged parties (see Lehmann and Gilson, 2015). Embedded research also usually involves a continuous flexible (re)negotiation of the research enterprise and relationships to build consensus and achieve synergies between actors across all relevant levels. In this way, embedded research is similar to the action research process – where changes or improvements are achieved thorough 'cycles of investigation, action, and reflection, while at the same time reporting it in a way that is useful to both to the project in hand and potentially to outsiders' (Costley et al., 2010: 88).

Negotiation towards longer-term partnerships: Embedded HPS research often requires a longer-term timeframe and intent. A particular embedded research project (or series of related projects) will commonly stretch out for several years, as it is unlikely that it can be properly negotiated or executed over a shorter period (e.g. the question of negotiation alone requires more time than in traditional health sciences research). An embedded approach usually emerges as a series of linked research and health systems strengthening activities, with overarching trusting relationships holding the activities together. For example, a long-term relationship between a specific academic department and a health system is usually built up over a series of projects, and as a result of continuous negotiation, dialogue, and the building of trust. What this means is that in embedded HPSR it is not only the short-term goals of the immediate research project that are negotiated (which is the perspective taken in much of the 'insider-research' literature), but also the longer-term objectives. Longer-term, more committed research relationships can foster further problem identification and future collaboration, can build research capacity within health systems to develop more resilient 'learning systems', and therefore might itself become a systems strengthening element. Thus while traditionally research projects might require negotiation on entry and exit, in embedded HPSR, negotiation is required continuously, and ideally should take a longer-term view of the relationships being built. Short-term research projects have the potential to damage relationships and create research fatigue in the health system. For example, in our experience, short-term contract evaluations, and thesis projects where students are inserted briefly into the system without careful mentoring have significant risk of damaging relationships for future HPSR.

Embedded research requires the careful analysis and negotiation of power: Sheikh et al. (2014) stress that all HPS researchers need to 'acknowledge and address questions around their own role and power as actors in the same system'. The role of power in health systems and in HPSR is of particular importance. In HPSR, power has an influence on what priorities are set, what questions can be asked, and what answers can be given (and absorbed). Power is a key influencer in the research negotiation described so far. An embedded researcher therefore needs to be sensitive to power issues (their own as well as the power of those with whom they engage), and have developed competencies to manage power differences or tensions that power differences often raise. In embedded HPSR, power is particularly visible in the negotiation of the problem and the question (whose priority is given precedence?); the negotiation of the research method (disciplinary or quantitative/qualitative negotiations?); power differentials between different types of health systems actors and decision-makers at different levels (for example, a manager's perspective is often problematically prioritized above that of a health worker or community member); or other socio-demographic imbalances.

Negotiation of information ownership and terms of co-production: An embedded approach requires researchers to take a step beyond the basic feedback processes usually required in research projects (where results are briefly reported back to stakeholders, often in the form of a publication). Instead, HPS researchers argue that in an embedded approach, processes that encourage the co-creation or co-production of knowledge are important (see Dunston et al., 2009; Goodyear-Smith et al., 2015; Jackson and Greenhalgh, 2015; Lehmann and Gilson, 2015). Co-production or co-creation is understood to be an ongoing collaborative process where researchers and end-users 'work together from the outset to frame relevant research questions, create research designs that map real-world environments, and commit to implementing the research and its findings in the broader health service community' (Jackson and Greenhalgh, 2015: 283).

It is argued that co-production helps reduce the gap between research, policy, and decision-making and encourages timely uptake of recommendations or interventions. But co-production also faces challenges such as power relations, differing perspectives on the issues at stake, and differing discursive skills in the different practice or research settings (see Pohl et al., 2010). Co-production is often discussed in relation to the basic issue of negotiation over publication authorship, where it manifests most visibly. However, co-production is much more complex than that, and is better understood as a part of the ongoing negotiation, and reflexivity – important to embedded HPSR. Reflexivity is 'the process of examining and recording the impact of researcher and intersubjective elements in research' (Freshwater, 2005: 311). In our experience, the heterogeneity of disciplinary perspectives within the team of health system partners and researchers, together with the influence of different organizational roles and experience at different levels of the health system, add to the complexity inherent in jointly making sense of findings. Furthermore, we have found that, as different research activities unfold over time, so the nature and practice of co-production has evolved. Negotiating shared accountability – for the research and its impact on the system – is also important. We like the idea from the WHO HPSR Strategy of health systems 'accomplices'.

> Embedding of research leads to greater accountability of evidence, both in terms of the evidence obtained and its use in decision-making . . . Policy-makers and researchers must not only work together as allies, they must work as accomplices and hold each other accountable for the impact – or lack thereof, of policies intended to improve the health of populations.
>
> (WHO, 2012: 20)

Embedded HPSR embraces context and complex 'adapting' systems

In HPSR, rigour is '. . . driven by engagement, contextualization, closeness to operational reality, attention to relevance, and reflexivity and the values that drive research practice' (Sheikh et al., 2014). Embedded research approaches are particularly apt for research that seeks to understand problems in context. Addressing social and economic embedded research (not specifically HPSR), Hirsch et al. (1990) note that embedded research is essentially characterized by the way it can take on rich empirical contexts, and equates embedded research with 'getting your hands dirty'.

It is widely acknowledged that health systems are not fixed constructs and that they change as we look at them; indeed, HPSR commonly acknowledges health systems to be 'complex adaptive systems'[5] (see WHO, 2012). Embedded approaches are particularly suited to research in complex adaptive systems. In fact, a common characteristic of an embedded approach is that it is research that expects the research context and research object to change during the research (which is one of the reasons fixed embedded research designs are rare). We encourage emerging HPS researchers to pay particular attention to the footnote and comment sections of HPS research articles, as it is commonly here that many of the clues are placed about the embedded experience. For example, in an article by Agyepong et al.

(2012), the footnotes describe the dual affiliations of most of the authors, and the fact that roles and affiliations changed during the research period (for example: 'Over the period of the ADHA saga explored here, IAA was a district director of health services and then a regional director'). Actors move and change within health systems – both practitioners and researchers – and this movement can have wide-ranging effects, from interrupting the research activity (as new relationships need to be negotiated) to changing the research findings entirely. Changes in personnel are only one small example of possible systems change facing embedded researchers. Sarah et al. (2002: 538) describe 'insider action research' as being non-linear, and involving a process where outcomes '. . . unfold and reveal themselves through cycles of action and reflection within a dynamic *context*' (our emphasis; see also Coghlan, 2007). We argue that while not all HPSR can or should be transformational action research (see Loewenson et al., 2014), an overarching embedded research approach can usefully adopt the cyclical process promoted in action research – a process of negotiation and reassessment of the system's context.

One of the main challenges when taking context and adaptation into account is that it then becomes difficult to generalize externally or transfer findings directly into other contexts. This form of generalizability is relevant in many areas of health sciences, but is less often found in HPSR, where analytic and or theoretical generalizability is increasingly being sought across different settings (see Gilson, 2012; Gilson et al., 2011). Certainly in embedded HPSR, it would be more common to rely on forms of analytic generalizability, or what Costley et al. call 'fuzzy generalizations':

> These are generalizations that arise from the particular research project and may have some general application in a similar context. Predicting easy solutions to diverse and complex issues involving a range of people is not realistic. Work based research may not transfer exactly to another situation, but it involves the application of research which has usefulness and application to a particular situation. It has usefulness to the community of practice and to the individual researcher, and it has the potential to generate theory. It embraces complexity and can be empowering and innovative, saving time and money by making improvements.
>
> (Costley et al., 2010: 3–4)

For example, conclusions drawn from embedded HPSR work might focus less on generalizing claims about the 'best solution' to a problem from one health system to the next, and might focus instead on generalizable lessons about 'how solutions were reached' for consideration in other systems.

Building and maintaining trust – research allies, trustworthy results, and ethics

We have addressed trust-building above but it is worth re-emphasizing. Hoffman et al. argue:

> . . . embedding health systems research as a core health systems function is about trust. For the successful translation of research to action, policymakers must trust the findings of researchers, and in turn, researchers must trust that

policymakers will not misuse (or abuse) their work. The benefits of such trust in developing, prioritizing and embedding health systems research also extend beyond health systems: such actions can be a key component of reform agendas promoting efficiency, good governance and accountability.

(Hoffman et al., 2012: 27)

We argue that an embedded HPSR approach is not possible without the establishment and maintenance of trust – at all levels. This is particularly important where co-production of knowledge is part of the approach used. In most insider-research guides, trust-building is focused mainly on the process of seeking endorsement from key stakeholders prior to when the research commences. In HPSR, it is also considered good practice for researchers to gain endorsement from the appropriate ethical and regulatory committees (usually in both the university and the local health system). We argue that in *embedded HPSR*, official endorsement is only a small part of a larger trust-building process between researchers and stakeholders. There is a large literature that addresses processes for building 'trustworthy and authentic' relationships in research (see Box 2.2), which argues this is a different way of thinking about the relationship between the researcher and the participants – requiring consideration of the effects of all research actions on the trust held between partners. Heshusius (1994) promotes an approach where relationship-building in research should be viewed as a methodology in and of itself. We endorse this view for an embedded approach to HPSR.

Building trusting relationships in embedded HPSR involves both pragmatic actions and relational competencies on the part of the embedded researchers. For example, it is understood that trust is built through cycles of collaborative reflection, action, and learning from each other, talking and listening within safe spaces where the value of each member is recognized (see Sheikh et al., 2014). Providing timely feedback and sharing information or data increases transparency and fosters trust. Trust is also built by openly communicating and developing a shared purpose. In our experience, poorly executed embedded research can equally result in broken trust, especially when expectations are raised and then not met. Because embedded research is often built on a web of long-term relationships, a single disagreement, if not managed appropriately, can destroy the specific embedded research project as well as the broader network of relationships linked to it. For example, as noted earlier, masters level students are sometimes sent out into health systems to conduct small thesis projects, usually inserted into a relational network that is in place between an academic institution and their health system. However, although they may have begun to master methodological competencies, inexperienced researchers are often not skilled or knowledgeable enough to be able to effectively negotiate, build trust or strengthen a more complex web of research relationships. They might intend to conduct embedded HPSR, but their degree structure is often more suited to rapid research, after which (if reminded) they might send the resulting thesis to key stakeholders several months later. This kind of research is valuable in its own right, but is not what we would consider 'embedded HPSR' (even if the student also happens to be a health worker or manager). Students' supervisors might need to play a more active role in research negotiation and act as the connector for the longer-term relationship.

Box 2.2:   Further reading resources

Action research (participatory action research)

See Loewenson et al. (2014)

Ethnography

See Albrecht et al. (2003), Cavanagh (2005)

Evaluation (insider evaluation, realist evaluation)

See Dahlberg and McCaig (2010)

Insider research (practitioner research, work-based learning/research, insider-led research, worker research, insider action research)

See Coghlan (2007), Costley et al. (2010), Reed and Procter (1995), Robson (2002), Workman (2007)

Implementation science (implementation research)

See Goodyear-Smith et al. (2015)

Research to practice (evidence-based policy/decision-making, policy process, research to policy, research utilization, knowledge transfer)

See Green and Bennett (2007), Koon et al. (2012)

Co-production of knowledge (co-creation, co-design, practice-research engagement, practitioners as writers)

See Jackson and Greenhalgh (2015), Lehmann and Gilson (2015)

Interdisciplinary research (inter-sectoral practice, intersectional research, collaboration practice, translation)

See Bammer (2013), Rigg (2011)

Reflective practice (coaching, mentorship, leadership training)

See Costley et al. (2010)

Learning systems (learning organizations)

See Garavan (1997), Watkins and Marsick (1993)

Trust-building in embedded HPS research might benefit from processes that establish 'safe spaces' – non-public spaces where research and findings can be negotiated without fear of exposure or reprisal (see, for example, the Executive Session approach; Moore and Hartmann, 1999). In both health and development sectors, it is acknowledged that negative findings often get hidden, and there is an urgent need to find effective processes where 'negative' implementation experiences, for example, can be shared in a non-threatening environment. We would argue that processes that establish 'safe spaces' are an important tool for HPS researchers intent on an embedded approach. In our experience, even critical research findings can eventually be made public if they emerge out of an authentic embedded process built on principles of co-creation and trust.

On the other hand, one of the tensions in embedded research is between the building of trusting relationships and the 'trustworthiness' of research findings. Cavanagh argues for an approach where: 'Thinking about validity in traditional terms of objectivity and lack of bias no longer applies. Rather, validity is seen in terms of building relationships based on trustworthiness and authenticity' (2005: 32). Gergen and Gergen (2000) argue that traditional efforts to discover and record the trustworthiness of research are being replaced by reflexivity and showing multiple perspective. In HPSR, there has been a significant push for new approaches to the assessment of rigour (see Gilson et al., 2011), although this remains challenging for HPS researchers emerging from a normative or positivist epistemology.

Again, we present embedded HPSR as an overarching approach rather than a specific tool or method. In practice, rigour requirements therefore remain those of the specific tool or method applied. However, we argue that in embedded HPSR, *additional rigour considerations need to be contemplated*. For example, the authenticity of the negotiation process, the balance of power in co-creation, and the trust held between research partners, might all be additional indicators of the 'trustworthiness' of the research findings.

We conclude this section with a comment about research ethics for embedded HPSR. We have noted in several places that an embedded approach raises specific ethical concerns. We reiterate that standard ethical processes and concerns apply – such as the appropriate protection of human subjects involved in research, and the reasons behind ensuring confidentiality and anonymity (which are often key concerns in HPSR). An embedded approach imposes additional ethical requirements on HPS researchers. For example, embedded researchers might also need to consider the *unintended consequences of their research on the health system*, which might extend beyond what ethical committees require. For example, in several of our previous research projects, even though anonymity of a district/facility was not required by the local partners or the ethics committees, it was jointly decided during the research that naming that district/facility might have a damaging ripple effect on the broader system. Ethical research engagement therefore also requires renegotiation and consideration of possible unintended consequences beyond the immediate scope of those particular research projects.

In a research approach where the researcher is inside the system and the focus is action on the system, a key ethical consideration is that the research *does not damage the health system in which it is based*. This goes beyond individual ethical protection, to an ethical protection and *consideration of risk to the health system*. In embedded HPSR, research *is* action – which carries a responsibility. Once embedded HPS researchers have negotiated their way into a position of trust within a health system, they will then usually face ethical considerations beyond those being taught in research methodology training, and also unlikely to be policed by ethical committees. We therefore need HPS researchers competent to negotiate the ethical considerations inherent in HPSR.

Goodyear-Smith et al. (2015) make an interesting observation about implementation research that utilizes participatory and co-design approaches (which is very close to what we are terming embedded research), and how this raises unique challenges for health sciences research ethics committees, which are more accustomed to dealing with traditional research approaches such as randomized controlled trials or ethnographies. They note that the flexibility and adaptation to context that is essential to co-created research, as well as the time necessary to build trust and co-create research questions, often has to be 'designed out' in order to pass through ethical review. They argue that ethics committees need to come to 'acknowledge and celebrate the diversity of research approaches . . . [and] ground rules should be established for co-design applications (e.g. how to judge when "consultation" or "engagement" becomes research) and communicated to committee members and stakeholders'. We argue that similar considerations need to be made by the ethics committees being asked to assess protocols utilizing an embedded HPSR approach.

In this section, we have outlined some characteristics and challenges common in an embedded approach to HPSR, and argue that the mechanics of embedded HPSR need to be more intentionally considered by practitioners and researchers in the field. Embedded HPSR is not an easy option, nor is it a tool that can be broken into clear steps. Rather, it is a complex approach requiring competencies and sensitivities for negotiation, collaboration, translation, trust-building, and reflexivity (as awareness of the context and awareness of self). Embedded research is critically important in HPSR, and we need to develop robust HPS researchers who are able to negotiate this complex world and wield this approach with confidence. In the next section, we present a brief case of embedded HPSR from one of our projects in South Africa, which depicts the characteristics outlined above.

Case study of embedded approaches in practice in Africa: the District Innovation and Action Learning for Health System Strengthening Project (DIAHLS)[6]

The DIAHLS project is a participatory embedded HPSR project that has been running in the Western Cape province of South Africa since 2010. It is built on a long-term partnership between the health departments of the City of Cape Town and the Provincial Government of the Western Cape and the Schools of Public Health at the Universities of Cape Town and the Western Cape, and is based in one health sub-district in Cape Town (the Mitchells Plain health sub-district). It was established as a collaboration between health managers and HPS researchers in Cape Town and much time was put into early negotiation between stakeholders (see intervention chronology in Table 2.2 and Box 2.3).

The initial research focus of DIAHLS-related activities was jointly decided as local governance and management processes, with specific concern for health system strengthening through joint learning. At the start of each new year research agenda, questions and activities are consistently (re)negotiated at a formal meeting that includes the sub-district managers. The focus and strategy of each linked strand of HPS *research* is also negotiated within the relevant group of health actors (and communicated to the broader group). In DIAHLS, the negotiated

Box 2.3:   Practical advice for embedded HPSR based on the DIAHLS experience

- Give enough time to project initiation and negotiation (at the start and then repeated at regular intervals)
- Begin with a longer-term commitment (ideally that can continue in some form even if dedicated funding does not) based on joint goals and objectives
- Build trust by being dependable and present, as well as being transparent about what different things each stakeholder is gaining from and giving to the collaboration
- Always provide rapid feedback (e.g. notes from meetings)
- Track the influence of context with respect to actors, policy change, and new initiatives that are impacting on the research setting
- Seek to provide new opportunities to the health system actors (e.g. let them know about conferences and new readings)
- Establish a process to handle staff/stakeholders cycling in and out of the system over time
- Establish a routine meeting process (initially with careful facilitation) that matches the pace and expectations of that particular group
- Explore and utilize existing tools for reflective practice and research within the embedded HPS research approach framing
- Include health service partners in reflective practice and in the interpretation of research findings
- Be respectful of each other, of different priorities, of different forms of 'research', and of different forms of evidence/knowledge
- Actively seek ways to acknowledge the contribution of health system partners through joint conference presentation, article writing
- Actively support the generation of research products that have value in the practitioner world such as reports to management meetings and policy briefs

focus has evolved over time towards a particular focus on understanding leadership and management in the sub-district, and how it is influenced by the district, provincial, and national structures, processes and policy environment in which it is located.

DIALHS has adopted a collaborative action learning approach based on Rigg's understanding of action research as 'a collective process for inquiring into and taking action on projects and practices within their complex, multi-agent contexts' (Rigg, 2011: 15). DIALHS aims to learn 'with' rather than 'about' health system actors (Lehmann and Gilson, 2015). For example, the sub-district management teams and facility managers are described as 'co-researchers of their own practice'. They contribute to generating data in facilitated dialogues and workshops, to iterative cycles of analysis, by making meaning of their experience and of the emerging findings, and also in joint authorship. The *action* component of action learning is understood within DIALHS to include reflective practice, the 'purposeful,

Table 2.2 Chronology of intervention and post-intervention analysis to strengthen the HAST programme

Who	What	Activities
Researchers	Document review	Review of minutes and proposals submitted to joint management meetings which identified problem
Task team	Collaborative reflective learning (1 meeting)	Problem analysis Document review to trace history of problem
Task team with all HAST programme staff	First workshop (2 days)	Established ground rules for working together cooperatively and respectfully Mapped out common vision for HAST Identified obstacles to implementing it Comparison of job descriptions HAST programme staff mapped onto organizational organograms
Task team	Collaborative reflective learning (2 meetings)	Recognized the highly relational nature of programme support Acknowledged that an interpersonal and inter-organizational relationship focus was required Developed principles for communication Mapped out organizational lines of communication related to authority, technical support, and line management onto the organizational organograms
Task team with sub-district management	Meeting	Confirmed sub-district programme managers' scope of decision-making
Task team	Collaborative work	Documented organizational lines of communication related to authority, technical support, and line management mapped onto the organograms Documented current roles and responsibilities of HAST programme staff
Task team with all HAST programme staff	Final workshop (½ day)	Shared approach to communicating and working together agreed Shared understanding of current roles and responsibilities agreed Documented principles of working together
Task team	Post-intervention analysis: collaborative reflective learning (2 meetings)	Reflection on the most significant changes post the intervention

Source: Scott et al. (2014).

critical analysis of knowledge and experience in order to achieve deeper meaning and understanding' (Mann et al., 2009: 596).

The main locus of research within DIALHS is at the sub-district level, specifically Mitchells Plain, a low-income community with 32% unemployment and 32% of the population living in informal housing (with related service delivery challenges). Although the entry point is the sub-district, DIAHLS is understood to be embedded *across all levels of the health system* – drawing in managers at facility, district, and provincial level, as well as managers within certain specialized sub-district functions (such as environmental health), into the process of learning and reflecting on learning. Table 2.3 lists the engagements of different actor groups at different levels within the research. This engagement across levels allows for a rich understanding of the health system's context, as well as a better understanding of how influences from different levels of the health system impact on leadership and management at the sub-district level.

In the first year of engagement (2010), two strands of collaborative activity were designed to develop a collective understanding of the sub-district context and to build relationships. The first was an exercise conducted with health service partners to map the sub-district planning and management processes in relation to the district and province. Members of the academic research team conducted interviews with key informants and undertook a document review of policy documents, annual plans, reports, and meetings across levels from the district to facility level. The process of data gathering led to rich descriptions – covering management structures, organograms, cycles of formal planning, supervision and performance management systems, monitoring and evaluation, information management, and funding flows. Further dialogue also led to strengthened relationships by valuing the tacit knowledge that those who work in the health system have about the system in which they work. Health service partners participated in making sense of the interconnected nature of formal and informal processes and the role of routine meetings in enacting their management responsibilities. Mapping lines of authority and communication (in relation to organograms) revealed questions that health service colleagues were grappling with and led to a richer understanding of management intentions and how these were, or were not, actualized in processes and practice. This activity generated collective questions and a sensitization to managerial issues that were to be traced over the long term. It also identified perceived constraints to decision-making, and identified where managers had influence and room to manoeuvre in innovation. For example, managers found that, while they were constrained by historic budgeting practices, there was a measure of flexibility that they could use between lines of expenditure.

The second strand was a joint activity with health services and community stakeholders to map health needs and resources within the community, which culminated in two workshops attended by civil society groups, health committees, sub-district and facility managers and staff. Participants spoke of the benefits of getting to know other health actors and learning more about health services and the communities served. It changed the way health staff saw the community, as they began to recognize the value of community assets. Local area action

Table 2.3 A team approach across a range of engagements

The team	Engagement
Academic researchers 2 principal investigators 3 senior researchers 2 post-doctoral researchers	Different members led and supported different activities Monthly meeting to reflect and plan (2010–present) 6-monthly discussion of literature (2010–2011) Annual full-day reflection on meta-learning (2010–present) Presentation at conferences (2012–present)
Health service colleagues in sub-district 2 sub-district managers as lead co-researchers	Annual reflection on managerial challenges in sub-district and negotiation of the research agenda (2010–present) Ongoing individual reflection accompanied by 2 principal investigators (2010–present) Presentations at conferences (2012, 2013, 2015, 2016) Reflection together and writing 3 articles (2013, 2014, 2015)
2 sub-district management teams (consisting of 5 and 7 members respectively in the two organizations, in addition to the sub-district managers)	Mapping management and planning processes (2010) Community profiling (2010) Supporting local area action groups (2011–2013) Meetings and workshops on supervision, management and monitoring and evaluation (M&E) (2013–2015) Mentored implementation of a new supervision strategy for the sub-district (2014) Sub-district management team mentoring (2016)
16 facility managers	Community profiling (2010) 8 coaching sessions on understanding self in workplace (2012) 4 reflective meetings on putting management theory into practice (2013) 3 reflective workshops on decision-making and information use (2013) Meetings and workshops on supervision, management and M&E (2013–2014) Focus group on team competencies and system capabilities to support teams (2015) Reflecting together and writing an article on managing absenteeism (2015, ongoing)
Environmental health practitioners	Mentored collective reflection on building management skills to address environmental health problems Reflecting together and writing an article on managing absenteeism (2015)

(Continued)

Table 2.3 continued

The team	Engagement
3 HIV/TB programme managers	6 reflective task team meetings Workshop series (2.5 days) Presentations at conference (2013, 2014) Reflecting together and writing an article on role of governance at implementation level (2014)
HIV/TB programme staff	Workshop series (2.5 days) to address conflict in HIV/AIDS and TB programme roles
4 facility managers as co-researchers in a sub-study on decision-making	Reflection on own practice through sets of interviews, development of mind maps (2012–2013) 3 presentations at a conference (2013) Reflecting on and writing an article on using information (2015) Exploring and writing an article on the ethics of embedded research (2016)
6 facility managers as participants in a sub-study understanding their transition in assuming a managerial identity	Journal writing and accompanied reflection (2011–2012)
Health service colleagues above sub-district 2 district health managers (incumbents of this position both changed during the course of the research and their successors took over in an ongoing relationship)	Annual and *ad hoc* meetings to negotiate the research approach and to make sense of findings

groups grew from this process to take forward local inter-sectoral action to address abuse and environmental hazard of illegal dumping of rubbish.

Ongoing attention to context across the course of the project (over five years) has been through dialogues with provincial, district, and sub-district managers and has meant that the policy and system influences have been observed over time, particularly as the local health system responds to changes at different levels.

Using a local management challenge to investigate larger system issues: HIV/TB programme roles

A strand of research activity within DIAHLS provided support to the HIV/AIDs, STD and TB (HAST) programme in the sub-district, and demonstrated clear benefits of an embedded HPSR approach – demonstrating trust-building, collective learning, and knowledge translation through co-creation processes. As a remnant of historic fragmentation, two organizations (respectively under the provincial and local government health authorities) work in the sub-district to provide primary health services, both partners in DIALHS. The introduction of a new management post and person in the provincial organization to support the HAST programme led to a rise in tensions between the two organizations. Because of the trust that had been built up in the DIALHS project through prolonged engagement, members of the research team were asked to support a locally constituted task team (five sub-district programme and operational managers drawn from the two organizations) to resolve conflict surrounding HAST-related roles and responsibilities in the sub-district.

The task team engaged in a process of reflective learning to develop a response (the intervention) to the problem, first as a small team and then in collaboration with all HAST programme staff in the sub-district. Moon's (1999) stages of reflective learning were used across the meetings and workshops to process: first 'noticing' and 'making sense' of the presenting problems, then 'making meaning [and] working with meaning' to a final stage of 'transformative learning' that could serve to guide their future practice.

The researchers traced the history of the creation of the post in a document review that they presented to the task team, who used this to reflect on how their understandings differed, and how expectations ought to be resolved. The larger two-day workshop for all HAST programme staff that followed then sought to understand the nature of the problem, the underlying reasons/drivers, and the way forward in addressing these from the perspectives of all HAST staff. The first workshop began by establishing ground rules for working together in the workshop setting and developing a joint vision for the HAST programme in the sub-district (both organizations had a strongly client-focused orientation).

The main obstacles in achieving this vision were clustered as: conflict in communication; lack of respect for the organizational lines of authority; conflict over access to information in the separate health information systems to be used for monitoring and planning; and inefficiencies with separated training. When it became evident that staff did not know one another's job descriptions and relative positions within their respective organizations, these were shared in the workshop setting, giving the programme staff within each organization an opportunity to comment and ask questions. Both programme managers and staff had assumed that the two organizations' structures and processes mirrored each other but, when the two organograms were mapped out, significant differences emerged that impacted on their understanding of lines of authority and acceptable lines of communication, as well as on policy implementation pathways.

Despite this learning, it was still not possible for the HAST programme managers and staff to plan a way forward to achieve their common vision. The HAST programme staff called instead for a set of standard operating procedures, approved by the district managers, to provide instruction on how to work together. They were seeking to establish rule-based relationships for working with colleagues in the partner organization and to be protected from uncertainty within these relationships by official organizational decrees.

The task team reconvened after the first workshop and, through a process of reflection, realized that HAST staff do their work with and through a wide range of stakeholders across sub-district departments and levels, and between the two organizations. In addition to mastery in technical HAST programme knowledge and skills, they required a strong set of relationship skills and a clear framework for and understanding of the relationships: 'I think we are now doing the real work. The relationships are key to how we do our work' (Programme manager, reflective task team meeting, 13 February 2012).

This insight became a guide to further work by the task team and informed the decision to strengthen support for collaborative relationships rather than defining rules for working and communicating. They recognized that a recorded description of current roles and responsibilities was useful in promoting a joint understanding of how work was done in the two organizations, but decided that communication between the HAST team and programme/operational managers, rather than a fixed agreement, was essential in maintaining the working relationship and ensuring that all responsibilities were covered. Also, they wanted roles and responsibilities to evolve organically within the two organizations in response to changing needs.

The following was decided upon: there could be open communication between HAST programme staff, unrestrained by lines of organizational authority; collaborative work in planning joint campaigns was desirable; and specialist support could be offered across organizational boundaries. Informal processes of operational planning were agreed and the Integrated Sub-district Management Team was identified as the appropriate local structure for formal information sharing and joint strategic planning.

The task team presented their collaborative learning back to a final workshop with all HAST programme staff. A participatory method, building on the ground rules established in the first workshop, was used to enable programme staff and managers from both organizations to develop jointly a set of principles for working together constructively. This participatory method modelled and reinforced the approach to respectful and proactive relationship-building. The principles supported the internalization of trust-based values (such as programme staff who are passionate, persistent, and respectful in how they work) and norms to govern communication and collaborative work (such as around information sharing and proactive, constructive problem-solving when obstacles arise). One programme manager later reflected that the value of these principles lay in their ability to 'neutralize power and hierarchy' in relations between staff who have to work together collaboratively (Programme manager, reflective task team meeting, 16 May 2012).

In the ensuing post-intervention reflection and analysis, three core cross-cutting features were identified as important to the success of the intervention: understanding and acting to make understood the differences between the two organizations; understanding and acting to support the highly relational nature of the work of HAST staff; and developing, modelling, and operationalizing a set of relational norms and values hinged on respect and valuing the ability to collaborate to deliver a client-focused service.

In learning together through a process of reflection, the researchers and programme managers were co-producing knowledge. This knowledge was immediately fed back into management practice in the sub-district to help resolve an immediate management challenge. The knowledge was also translated, 'packaged', and made available to others in the health policy and system community through a poster presentation at a global conference (presented by one of the sub-district managers) and a jointly authored article (see Elloker et al., 2013). It also became part of the learning around the centrality of relationships and values in dialogue with the sub-district and district managers, and served as a graphic case study and reminder of this learning. Eight months after the last intervention workshop, evidence of the effect of the learning in practice was identified in other programmes in the sub-district. For example, the greater understanding of organizational differences and needs had enabled nurses who had been trained in child health to be placed in the partner organization for a period of experience and mentorship. It was noted that: 'working together [in HAST] had spilled over into other areas' and 'allowed for greater collaboration between the two organizations' (Reflective task team meeting, 16 April 2013).

Notes

1 We acknowledge that in HPSR it is often not easy or useful to make a distinction between practitioners and researchers. Practitioners are often researching, and researchers often practise in a variety of ways. We use these terms, but acknowledge that they are only useful to a certain extent and then become barriers to understanding just how complex and fluid health systems actors are. There are multiple examples of researching-practitioners, as well as academics or professional researchers routinely working from within the health system. We also are reminded that 'research' as we use it here should not only be understood as formal published (peer-reviewed) reports, but also includes routine and experiential learning.
2 In flexible design the research strategy evolves during data collection and often involves non-numerical data, while fixed designs uses strategies that are 'fixed' and specified before the data collection stage (Robson, 2002).
3 It is interesting that barely any of this literature is focused on low- and middle-income countries (LMIC) settings or examples, which is where HPSR is often focused. It might be useful to reflect on whether insider-research experiences are different in low- and middle-income countries (LMIC) settings.
4 In our experience, not a lot of teaching develops such competencies within HPSR – and this is an area for further development.
5 'Complex adaptive systems are described as such because in addition to being comprised of many interacting components, they have the capability to self-organize, adapt or learn from experience' (Paina and Peters, 2012: 367).
6 Authorship note: V.S. and L.G. are part of the DIAHLS case example described in the second part of this chapter, and have published about it (with other colleagues) more extensively elsewhere (see Cleary et al., 2014; Elloker et al., 2013; Gilson et al., 2014; Lehmann and Gilson, 2015).

References

Agyepong, I.A., Kodua, A., Adjei, S. and Adam, T. (2012) When 'solutions of yesterday become problems of today': crisis-ridden decision making in a complex adaptive system (CAS) – the Additional Duty Hours Allowance in Ghana, *Health Policy and Planning*, 27 (suppl. 4): iv20–iv31.

Albrecht, G.L., Fitzpatrick, R. and Scrimshaw, S.C. (2003) *The Handbook of Social Studies in Health and Medicine*. London: Sage.

Bammer, G. (2013) *Disciplining Interdisciplinarity: Integration and implementation sciences for researching complex real-world problems*. Canberra, ACT: ANU E Press.

Blanchet, K. and James, P. (2012) How to do (or not to do). . .a social network analysis in health systems research, *Health Policy and Planning*, 27 (5): 438–46.

Caffrey, L., Wolfe, C. and McKevitt, C. (2016) Embedding research in health systems: lessons from complexity theory, *Health Research Policy and Systems*, 14 (1): 1–9.

Cavanagh, T. (2005) Constructing ethnographic relationships: reflections on key issues and struggles in the field, *Waikato Journal of Education*, 11 (1): 27–42.

CHESAI (Collaboration for Health Systems Anaysis and Innovation) (2016) Retrieved from: www.chesai.org [accessed 1 April 2016].

Cleary, S.M., Schaay, N., Botes, E., Figlan, N., Lehmann, U. and Gilson, L. (2014) Re-imagining community participation at the district level: lessons from the DIALHS collaboration, in *South African Health Review 2014/15* (pp. 151–61). Durban: Health Systems Trust.

Coghlan, D. (2007) Insider action research: opportunities and challenges, *Management Research News*, 30 (5): 335–43.

Cook, K.S., Emerson, R.M., Gillmore, M.R. and Yamagishi, T. (1983) The distribution of power in exchange networks: theory and experimental results, *American Journal of Sociology*, 89 (2): 275–305.

Costley, C. and Gibbs, P. (2006) Researching others: care as an ethic for practitioner researchers, *Studies in Higher Education*, 31 (1): 89–98.

Costley, C., Elliott, G.C. and Gibbs, P. (2010) *Doing Work Based Research: Approaches to enquiry for insider-researchers*. London: Sage.

Dacin, M.T., Ventresca, M.J. and Beal, B.D. (1999) The embeddedness of organizations: dialogue and directions, *Journal of Management*, 25 (3): 317–56.

Dahlberg, L. and McCaig, C. (2010) *Practical Research and Evaluation: A start-to-finish guide for practitioners*. London: Sage.

Dunston, R., Lee, A., Boud, D., Brodie, P. and Chiarella, M. (2009) Co-production and health system reform – from re-imagining to re-making, *Australian Journal of Public Administration*, 68 (1): 39–52.

Elloker, S., Olckers, P., Gilson, L. and Lehmann, U. (2012) Crises, routines and innovations: the complexities and possibilities of sub-district management – leadership and governance, in *South African Health Review 2012* (pp. 161–73). Durban: Health Systems Trust.

Freshwater, D. (2005) Writing, rigour and reflexivity in nursing research, *Journal of Research in Nursing*, 10 (3): 311–15.

Garavan, T. (1997) The learning organization: a review and evaluation, *The Learning Organization*, 4 (1): 18–29.

George, A. (2009) 'By papers and pens, you can only do so much': views about accountability and human resource management from Indian government health administrators and workers, *International Journal of Health Planning and Management*, 24 (3): 205–24.

Gergen, M.M. and Gergen, K.J. (2000) Qualitative inquiry: tensions and transformations, in N.K. Denzin and Y.S. Lincoln (eds.) *Handbook of Qualitative Research*, 2nd edn. (pp. 1025–46). Thousand Oaks, CA: Sage.

Ghaffar, A., Langlois, E.V., Rasanathan, K., Peterson, S., Adedokun, L. and Tran, N.T. (2017) Strengthening health systems through embedded research, *Bulletin of the World Health Organization*, 95: 87.

Gilson, L. (ed.) (2012) *Health Policy and Systems Research: A methodology reader*. Geneva: Alliance for Health Policy and Systems Research/World Health Organization.

Gilson, L., Hanson, K., Sheikh, K., Agyepong, I.A., Ssengooba, F. and Bennett, S. (2011) Building the field of health policy and systems research: social science matters, *PLoS Medicine*, 8 (8): e1001079.

Gilson, L., Elloker, S., Olckers, P. and Lehmann, U. (2014) Advancing the application of systems thinking in health: South African examples of a leadership of sensemaking for primary health care, *Health Research Policy and Systems*, 12: 30 [doi: 10.1186/1478-4505-12-30].

Goodyear-Smith, F., Jackson, C. and Greenhalgh, T. (2015) Co-design and implementation research: challenges and solutions for ethics committees, *BMC Medical Ethics*, 16: 78 [doi: 10.1186/s12910-015-0072-2].

Green, A. and Bennett S. (eds.) (2007) *Sound Choices: Enhancing capacity for evidence-informed health policy*. Geneva: World Health Organization.

Heshusius, L. (1994) Freeing ourselves from objectivity: managing subjectivity or turning toward a participatory mode of consciousness?, *Educational Researcher*, 23 (3): 15–22.

Hirsch, P., Michaels, S. and Friedman, R. (1990) Clean models vs. dirty hands: why economics is different from sociology, in S. Zukin and P. DiMaggio (eds.) *Structures of Capital: The social organization of the economy* (pp. 39–56). Cambridge: Cambridge University Press.

Hoffman, S.J., Röttingen, J.-A., Bennett, S., Lavis, J.N., Edge, J.S. and Frenk, J. (2012) *A Review of Conceptual Barriers and Opportunities Facing Health Systems Research to Inform a Strategy from the World Health Organization*. Geneva: Alliance for Health Policy and Systems Research/WHO.

Jackson, C.L. and Greenhalgh, T. (2015) Co-creation: a new approach to optimising research impact?, *Medical Journal of Australia*, 203 (7): 283–4.

Khresheh, R. and Barclay, L. (2007) Practice-research engagement (PRE): Jordanian experience in three Ministry of Health hospitals, *Action Research*, 5 (2): 123–38.

Koduah, A., van Dijk, H. and Agyepong, I.A. (2015) The role of policy actors and contextual factors in policy agenda setting and formulation: maternal fee exemption policies in Ghana over four and a half decades, *Health Research Policy and Systems*, 13: 27 [doi: 10.1186/s12961-015-0016-9].

Koon, A., Nambiar, D. and Rao, D.R. (2012) *Embedding of Research into Decision-making Processes*. Geneva: Alliance for Health Policy and Systems Research/World Health Organization.

Lehmann, U. and Gilson, L. (2015) Action learning for health system governance: the reward and challenge of co-production, *Health Policy and Planning*, 30 (8): 957–63.

Loewenson, R., Laurell, A.C., Hogstedt, C., D'Ambruoso, L. and Shroff, Z. (2014) *Participatory Action Research in Health Systems: A methods reader*. Nairobi: Regional Network for Equity in Health in East and Southern Africa.

Mann, K., Gordon, J. and MacLeod, A. (2009) Reflection and reflective practice in health professions education: a systematic review, *Advances in Health Sciences Education*, 14 (4): 595–621.

Moon, J. (1999) *A Handbook of Reflective and Experiential Learning*. London: Routledge.

Moore, M.H. and Hartmann, F.X. (1999) *On the theory and practice of 'Executive Sessions'*. Unpublished manuscript [available at: https://scholar.google.fr/scholar?q=On+the+theory+and+practice+of+%E2%80%98Executive+Sessions%E2%80%99.+Unpublished+manuscript&hl=en&as_sdt=0&as_vis=1&oi=scholart&sa=X&ved=0ahUKEwiy2o-BwYnVAhVHNhoKHePKAN0QgQMIJjAA; accessed 14 July 2017].

Nee, V. and Ingram, P. (1998) Embeddedness and beyond: institutions, exchange, and social structure, in M.C. Brinton and V. Nee (eds.) *The New Institutionalism in Sociology* (pp. 19–45). New York: Russell Sage Foundation.

Paina, L. and Peters, D.H. (2012) Understanding pathways for scaling up health services through the lens of complex adaptive systems, *Health Policy and Planning*, 27 (5): 365–73.

Pohl, C., Rist, S., Zimmermann, A., Fry, P., Gurung, G.S., Schneider, F. et al. (2010) Researchers' roles in knowledge co-production: experience from sustainability research in Kenya, Switzerland, Bolivia and Nepal, *Science and Public Policy*, 37 (4): 267–81.

Reed, J. and Procter, S. (eds.) (1995) *Practitioner Research in Health Care: The inside story*. London: Chapman & Hall.

RESYST (Resilient and Responsive Health Systems) (2016) Retrieved from: http://resyst.lshtm.ac.uk/ [accessed 10 April 2016].

Rigg, C. (2011) Systemic action and learning in public service, *Action Learning: Research and Practice*, 8 (1): 15–26.

Robson, C. (2002) *Real World Research*, 2nd edn. Oxford: Blackwell Publishing.

Sarah, R., Haslett, T., Molineux, J., Olsen, J., Stephens, J., Tepe, S. et al. (2002) Business action research in practice: a strategic conversation about conducting action research in business organizations, *Systemic Practice and Action Research*, 15 (6): 535–46.

Scott, V., Schaay, N., Olckers, P., Nqana, N., Lehmann, U. and Gilson, L. (2014) Exploring the nature of governance at the level of implementation for health system strengthening: the DIALHS experience, *Health Policy and Planning*, 29 (suppl. 2): ii59–ii70.

Sheikh, K., George, A. and Gilson, L. (2014) People-centred science: strengthening the practice of health policy and systems research, *Health Research Policy and Systems*, 12: 19 [doi: 10.1186/1478-4505-12-19].

Watkins, K.E. and Marsick, V.J. (1993) *Sculpting the Learning Organization: Lessons in the art and science of systemic change*. San Francisco, CA: ERIC.

World Health Organization (WHO) (2012) *Changing Mindsets: Strategy on health policy and systems research*. Geneva: WHO.

Workman, B. (2007) Casing the joint: explorations by the insider-researcher preparing for work-based projects, *Journal of Workplace Learning*, 19 (3): 146–60.

3 Boundary critique: an approach for framing methodological design

Martin Reynolds and Helen Wilding

Introduction

Boundary critique is not a research method, or indeed a methodology. Boundary critique is an underlying reflective systems approach to methodological design. It is a 'systems' approach because it deals explicitly with boundaries and boundary judgements – a core feature of any systems approach to research (Ulrich, 2001). Making boundary judgements contributes to exclusion and marginalization (of people and ideas) and so it is necessary to subject boundary judgements to critical reflection, debate, and justification – which, in sum, is what boundary critique is all about (Ulrich, 2003). Boundary critique can help in (re)formulating research questions and aims in order to make them more relevant to practical needs. It can also help draw out the implications of research for different stakeholder audiences, including policy decision-makers at different levels, and civic society. Boundary critique can be adapted for use by any user with any one method or collection of methods (conventional and/ or systems-based) comprising a specific methodology, in any research context where there is complexity. The core idea behind boundary critique is that it fosters a conversation between empirical judgements ('facts') and normative judgements ('values'). Boundary critique provides a way to reframe conversations about where we are, where we ought to be going, and what makes that choice the right way to go. It therefore speaks to concerns of any practical research:

> Ensuring that research is translated into practical outcomes, and in turn ensuring that research priorities and activities are informed by practitioner experience, is a recognised priority in sectors such as medicine . . .
>
> (Real KM, 2015)

This statement from a knowledge management agency implies that research and practical experience are separate activities: one dealing with understanding the factual world (research) and the other dealing with the world of values (practical experience). In reality, of course, the two worlds are interchangeable, but the relative (pre)dominance of 'facts' and 'values' is variable depending on different approaches. The popularity of contemporary evidence-based practice (EBP) approaches in health care have a long and well-documented history (cf. Guyatt and Rennie, 2002; Melnyk and Fineout-Overholt, 2011; Walshe and Rundall, 2001), and would tend to prioritize the factual world over the world of values. It comprises a conventional linear mechanistic understanding of 'research' informing 'practice'; an understanding described elsewhere in the arena of public adminis-tration as the 'Received View' (Cook and Wagenaar, 2012).

In the arena of health care, the importance of values is paramount (Porter, 2010), particularly in setting priorities of health systems research in low-income countries (Nuyens, 2007). An alternative approach to the Received View is what we might call a praxis-oriented approach. Here, the continual integral dynamic between research and practice is acknowledged, recognizing also that research is essentially value-driven. Praxis can be described as practice-informed-theory and/or theory-informed-practice; or alternatively, ideas-in-action or thinking-in-practice. In praxis, the activities of research and practice are not seen as an either/or dual-ism, but rather regarded as a continual both/and duality. One significant expres-sion of a praxis-oriented approach in the systems thinking tradition is Werner Ulrich's notion of boundary critique: an 'eternal triangle' of interdependence between judgements of fact and value judgements, mediated through boundary judgements (Ulrich, 1996a, 1998, 2000).

The first part of this chapter describes boundary critique and a particular manifes-tation of it – the systems thinking in practice (STiP) heuristic, developed by systems practitioners at the UK-based Open University. STiP provides a framework for using multiple methods or techniques – systems-based or otherwise – through a sequence of activities involving: (1) understanding interrelationships, (2) engaging with multi-ple perspectives, and (3) reflecting on boundary judgements. Two sets of systems tools have been found particularly helpful in these activities, as experienced by the authors: tools associated with soft systems methodology (SSM) and critical systems heuristics (CSH). These are briefly outlined. The second part of the chapter describes an application of the STiP heuristic in developing systems for health partnerships. While the case study is situated in a UK context – specifically Newcastle upon Tyne, in the North East of England – the wider context of enabling partnerships through the application of boundary critique is one that has universal relevance.

Boundary critique and systems praxis

Making boundary judgements

Systemic intervention for public health has been described as 'purposeful action by an agent to create change in relation to reflection upon boundaries' (Midgley, 2006: 467). Making boundary judgements constitutes the core of systems think-ing. Whenever the term 'system' is used – whether through common language understanding of, say, the legal system, economic system or health system, etc., or through specific systems approaches to inquiry (system dynamics, complex adaptive systems, soft systems methodology, etc.) – there are implicit or explicit boundaries invoked. Boundaries mark out the map (systems of interest) from the territory (messy or 'wicked' situations of interest).

Using systems thinking for effecting change in a situation requires reflecting upon boundaries, both in terms of systems being evaluated and systems for design. Boundary critique is a term used by both Gerald Midgley and Werner Ulrich, both drawing on the work of systems philosopher West Churchman (1970), to describe the way in which boundary judgements might be used as a reflective device. Here, we will focus on Ulrich's ideas. For Ulrich (1996a), boundary judgements reflect both value judgements – in relation to ethical issues of marginalization (cf. Midgley, 2006) – and factual judgements – particularly relevant in relation to the importance of evidence-based approaches to health systems research.[1]

The juxtaposing of factual judgements with value judgements represents a divide of two cultures in contemporary health systems thinking (evidence-based and values-based), which in turn signals the need for better integrated modelling (Lang and Rayner, 2012) and more politically engaged practice (e.g. van Olmen et al., 2012) in health systems research. Forging a meaningful conversation between facts and values lies at the heart of boundary critique.

Boundary critique involves systemic triangulation, what Ulrich calls an 'eternal triangle': an interdependence between judgements of fact, value judgements, and boundary judgements (Ulrich, 1998: 6; 2000: 252):

> Thinking through the triangle means to consider each of its corners in the light of the other two. For example, what new facts become relevant if we expand the boundaries of the reference system or modify our value judgments? How do our valuations look if we consider new facts that refer to a modified reference system? In what way may our reference system fail to do justice to the perspective of different stakeholder groups? Any claim that does not reflect on the underpinning 'triangle' of boundary judgments, judgments of facts, and value judgments, risks claiming too much, by not disclosing its built-in selectivity.
>
> (Ulrich, 2003: 334)

Ulrich provided an original figurative triangle of facts, values, and boundary judgements (Ulrich, 2000: 252) that has been substantially developed and revised here by one of the co-authors, Martin Reynolds (Figure 3.1). The three poles of the triangle represent: (1) the reality of the situation of interest being mapped for purposes of understanding and design for change – for example, the entities, stakeholders, and conflicts of a messy health-related set of issues; (2) the values among key stakeholders responsible for bringing about change in the situation; and (3) the boundaries of systems used for effecting change. Such boundaries are always going to be partial. All systems are necessarily partial – or selective – in the dual sense of (i) representing only a section rather than the whole of the total universe of considerations, and (ii) serving some parties – or interests – better than others (Ulrich, 2002: 41). In other words, no proposal, no decision, no action, no methodology, no approach, no system can get a total grip on the situation (as a framework for understanding) or get it right for everyone (as a framework for practice) (Reynolds, 2008).

Forging a meaningful conversation between facts and values through reflecting on boundaries lies at the heart of praxis – thinking in practice. Boundary critique provides a set of principles for enacting systemic praxis. Systems thinking in practice (STiP) translates boundary critique principles into a more workable approach of praxis.

The STiP approach

Systems thinking in practice (STiP) is the namesake of a postgraduate programme of study developed by a group of systems practitioners at The Open

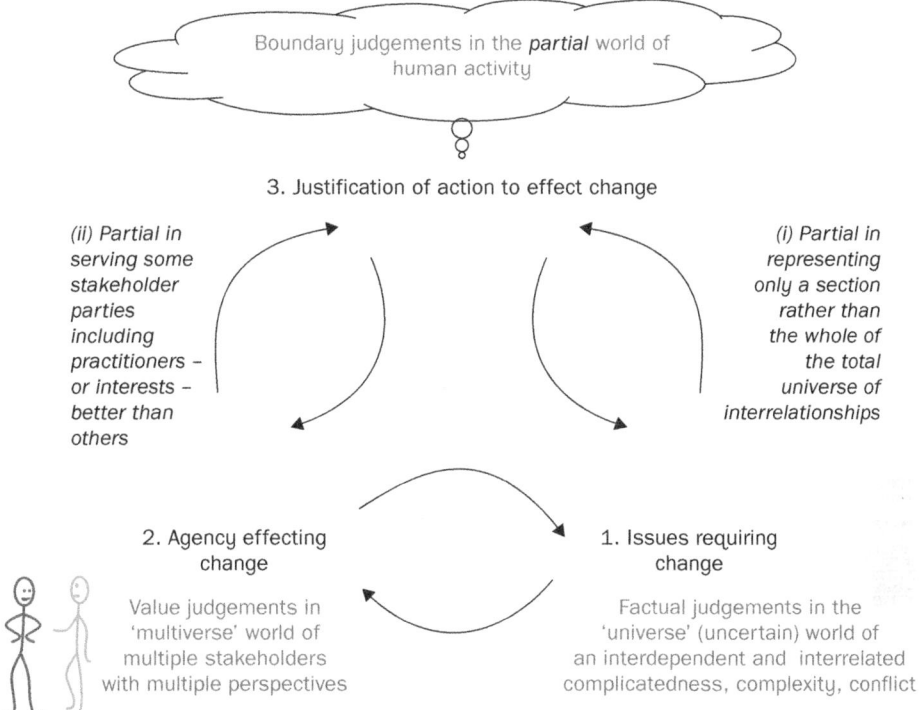

Figure 3.1 Systems dynamics
Source: adapted from Reynolds (2011). © Martin Reynolds, 2017.

University in the UK (Blackmore et al., 2014; Open University, 2012; Reynolds, 2014a). The two core modules provide a steer in applying systems ideas to students' professional work practices to develop systemic praxis.

The heuristic (Figure 3.2) can be explained in terms of three core activities, associated respectively with (and derived from) the three axes of boundary critique (facts, values, and boundaries):

i.   *Understanding* interrelationships – *capturing* complicatedness (inter-dependencies), complexity (perspectives), and conflict (in boundaries);
ii. *Engaging* with multiple perspectives – *designing* systems for health using systems as proxy to perspectives (systems of interest); and
iii. *Reflecting* on boundary judgements – the possibilities and challenges – used in (i) the capturing of interrelationships and (ii) design of new proposed systems of interest.

The numbers 1–3 on Figure 3.2 correspond directly with numbers 1–3 relating to boundary critique (Figure 3.1). Systems here are regarded explicitly as conceptual constructs – tools (item 3) used for understanding situations (item 1) and engaging with different perspectives (item 2). Different tools from different traditions (including but not only 'systems' traditions) can be adapted and used for each of the three core activities of STiP, depending on: (1) the context of use (including other stakeholders' involvement) and (2) the experiences and skills of

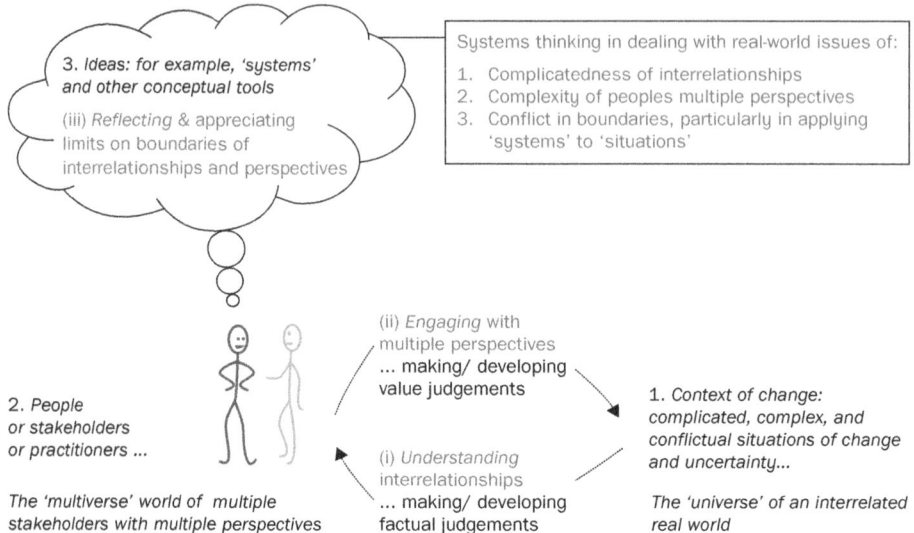

Figure 3.2 Systems thinking in practice heuristic

Source: Reynolds (2014), adapted from Open University (2012). © Martin Reynolds, 2017.

the user(s). The tools chosen below are those used in the particular case study reported on in the next section – developing partnerships for health.

(Systems) tools 1: simple PQR (What?/How?/Why?) systems

Simple systems originate from the tradition of soft systems methodology (SSM) developed by Peter Checkland (Checkland, 1981, 2011; Checkland and Scholes, 1990), and further elaborated upon by Rosalind Armson (2011).

The shift in representing reality from relatively unbounded 'rich pictures' (a compilation of drawings, pictures, symbols, and text that represent a particular situation or issue from the viewpoint(s) of the person (or people) who drew them) to bounded systems can occur through a series of stages. Each stage successively offers a higher level of granularity. At the lowest level of granularity, systems can be represented as expressions of purpose. Open University practitioners have used the term 'snappy' systems to denote such systems, where quick-fire purposes might be aligned with any one particular theme of interest. For example, the following snappy systems can be associated with two themes of health systems research:

Theme 1: primary health care in low-income countries

A system to:

- mobilize support from middle-income and high-income countries
- train potential local nurses and doctors and auxiliary staff
- develop infrastructure support for primary clinics
- monitor relative impact of primary and secondary healthcare provision
- etc.

Theme 2: outbreaks of infectious diseases

A system to:

• provide vaccination
• invest in preventative measures supporting hygiene
• identify sources of infectious disease
• control the use of antibiotics
• etc.

Each snappy system represents a particular perspective in terms of the purpose associated with a particular theme. The next level of granularity would involve unfolding any one purpose in to a simple system in terms of not only *what* it does/should do, but also *how* it might be undertaken, and *why* it's being carried out.

Rosalind Armson (2011: 213–38) provides a simple template for defining a human-activity system – using the idea of a purpose and sub-systems of activity: a system to do *What* by means of *How* in order to contribute to achieving *Why*. Examples might be:

• A system to (What?) provide localized health care training by means of (How?) government investment in order to (Why?) improve primary healthcare provision.
• A system to control the use of antibiotics by means of advising doctors and patients appropriately about risks in order to provide better, more efficient overall healthcare service.

Each system has only one What, the primary activity that defines its purpose, and one Why. There may be several types of How associated with the system but they must be coherent in marrying the What with the Why. So a system to control the use of antibiotics might have How sub-systems associated with, say, reducing antibiotic availability or policing stocks of antibiotics, but such measures might not be most appropriate for addressing the Why: to reduce the occurrence of drug-resistance among pathogenic organisms.

Simple PQR systems can be used to clarify actions either perceived to be practised in reality or as design models for transforming the reality. However, as Armson suggests: 'constructing a system definition is deceptively simple but rigorous and challenging [. . .] The challenge is "to say precisely what you mean and mean precisely what you say" by expressing the essential transformation in its clearest form' (2011: 36).

A further level of granularity in systems thinking is offered by unfolding a simple system even further through using critical systems heuristics as a reference system.

(Systems) tools 2: critical systems heuristics (CSH)

Critical systems heuristics provides a reference system that is an elaboration of a simple system. CSH was initiated by Werner Ulrich (1983), based on the work of Churchman (1979) and developed further with one of the co-authors, Martin

Reynolds (Ulrich and Reynolds, 2010). CSH provides a reference system circumscribed by four sources of influence (elaborated below by Reynolds, 2014b):

1. *Values and motivations* built into our views of situations and efforts to 'improve' them (who gets what?).
2. *Power structures* influencing what is considered a 'problem' and what may be done about it (who owns what?).
3. *The knowledge basis* defining what counts as relevant information and skills (who does what?).
4. *The moral basis* on which we expect third parties (i.e. people and/or non-human nature not involved yet in some way concerned) to bear the consequences of what we do, or fail to do, about the situations in question (who gets affected by what some people get?).

In CSH, these four dimensions of a complex situation are called *sources of influence*. The first three sources – *motivation, control*, and *knowledge* – determine the coordinates/dynamics associated with those 'involved' with the system; the final source – *legitimacy* – represents the context in which the system operates (see column 'sources of influence' in Table 3.1).

Each of the four sources of influence have three bounded questions regarding who the stakeholders might be, what's at stake, and what might be the particular stakeholding issues or key problems associated with the particular stakeholder group. Thus there are a total of twelve boundary judgements to be made regarding any situation being examined. The twelve corresponding CSH questions can be asked in either an analytical 'is' mode or a more design-led normative 'ought' mode. Table 3.1 outlines the 12 boundary judgements associated with CSH.

The CSH template (involving questions CSH q1–12) therefore provides a more defined 'rich picture' of a situation, and the vehicle for developing a more elaborated systems model. Thus the 'what' (P) in PQR corresponds directly with the 'purpose' question (CSH q2) in the CSH reference system. The 'why' (R) in PQR corresponds with the 'emancipation' question (CSH q11), which signals not only the worldview underpinning the system in question but also the worldview being challenged through the implementation of the system. The 'how' (Q) component of PQR is associated with all the other elements of the reference system (CSH q1, 3, 6–11, 12), representing the agency required to effect the purpose (stakeholders, stakes, and stakeholding issues).

Collectively, CSH questions provide parameters of a wider political economy of a situation (cf. Reynolds, 2014b), which arguably chimes well with the 'ecological' model called for in public health thinking in the twenty-first century (Lang and Rayner, 2012).

Like all systems tools, the CSH reference system can be used and adapted for different needs. For a systematic in-depth description of how CSH might be used either generically or on issues relating to environmental planning, readers are referred to Ulrich (1996b), Reynolds (2007) and Ulrich and Reynolds (2010). For in-depth use in case studies relating specifically to health interventions, there are few citations available. Ulrich himself uses a comprehensive case study on health systems planning in Puget Sound, Washington State, USA, to illustrate the

Table 3.1 Boundary judgements as questions relating to CSH

	Boundary judgements informing a system of interest S, *where S may represent an intervention such as a policy, programme or project*		
Sources of influence	**Stakeholders**	**Stakes** *(specific interests)*	**Stakeholding issues** *(key problems)*
Who gets what? Sources of **motivation**	1. *Beneficiaries* Intended clients or customers of S?	2. *Purpose* Key objective of S?	3. *Measure of success* S's measure of improvement?
Who owns what? Sources of **control**	4. *Decision-makers* Those in command of resources necessary to enable S?	5. *Resources* Conditions of success for S – relevant components ('capital') to secure improvement?	6. *Decision environment* (accountability) Conditions of success outside the control of the decision-maker for S?
Who does what? Sources of **knowledge**	7. *Experts* Those providing relevant knowledge and skill for enabling S?	8. *Expertise* Relevant knowledge and skills supporting S?	9. *Guarantor* (assurances) Promise or guarantee of successful implementation of S?
Who gets affected by what some people get? Sources of **legitimacy**	10. *Witness* (victims) Those representing the interests of those negatively affected by but not involved with S?	11. *Emancipation* (marginalization) Constraints on the interests of those negatively affected to have expression and freedom from the worldview of S?	12. *Worldview* (political space) Opportunities available for reconciling contrasting worldviews giving meaning to improvement in S?

The 'involved' (rows 1–9), The 'affected' (rows 10–12)

Source: Adapted from Ulrich and Reynolds (2010) and Reynolds (2014a). © Martin Reynolds, 2017.

actual and ideal modes of CSH (Ulrich, 1983: 372–417). Midgley has also used CSH questions in a number of health-related inquiries, including examining the mental health problems of ex-prisoners, and the needs of older people in sheltered housing (discussion and further citations in Midgley, 2006).

It is difficult to gauge the extent to which ideas of CSH are known and/or adapted for use among health practitioners.[2] Using CSH in its entirety can be time-consuming, particularly for busy professionals who need to deal with issues quickly. But there are ways of using CSH more succinctly. Box 3.1 illustrates a quick turnaround understanding of CSH parameters by focusing principally on the stakeholding questions (CSH q3, 6, 9 and 12). The CSH-lite approach provides a thumbnail sketch of a situation of interest.

CSH can be adapted for use across different levels (personal action planning to national/global systems design) and in different contexts (development,

---

Box 3.1:   CSH-lite exploration

System of interest: 'a system to control the use of antibiotics'

The measures of success (CSH q3) can in the long term be associated with a reduction in the occurrence of drug resistance in pathogens and/or the reduction in costs of research towards developing new antibiotics, although a more immediate measure might be the reduction in antibiotic prescriptions. A key factor with attempts at changing existing professional practice is to keep check of other workload practices (CSH q6) that frontline health professionals are engaged with. Such professionals may possibly have limited capacity to focus on revising traditional practices of advice and assurance to patients, but instead revert to the false guarantor of a quick-fix solution signalled by antibiotic prescription (CSHq9).

The proposed system would therefore need space to challenge the existing system arguably dominated by corporate pharmaceutical interests profiteering in the supply of research and development for new antibiotics, coupled with a context where professional healthcare providers/practitioners have very limited time for conversations with citizen-users of healthcare provision (CSHq12).

---

business, education, health). In terms of the STiP heuristic, CSH provides prompts towards:

- Understanding interrelationships: 4 sources of influence × 3 issues of stake-holding.
- Engaging with multiple perspectives: 4 stakeholder groups/involved vs affected/ideal modelling.
- Reflecting on boundaries: comparing the 'ought' with the 'is' world of reality; checking on incidences of reductionism (not capturing relevant interrelationships), dogmatism (not engaging appropriately with different perspectives), and managerialism (checking false claims towards holism and pluralism).

Case study: systemic praxis in developing partnerships for health

Introduction and overview of partnership working for health

As the Tallinn Charter (WHO, 2008) highlights, health systems are made up of an ensemble of organizations and institutions. Therefore, any attempt to work to improve health systems necessarily involves collaborations or partnerships. Partnerships are seen as a way of coordinating the activities of a range of public, private, and non-governmental organizations towards a common aim. They are conceived as an organizational form in their own right and are often given concrete expression through the creation of a governance body such as a Board (Lowndes and Skelcher, 1998). Informal and formal partnerships form to address many different health concerns in a range of local, national, and international contexts, from developing primary healthcare provision in a village in a low-income country to implementing a health-in-all-policies approach to global health. In effect, partnership working offers the opportunity for organizations and institutions to work together as a system. However, different stakeholders will have different experiences of what that 'system' *is*: different visions of what it *ought* to be; different views of the role of entities that give expression to that partnership; and different proposals with regards to the action that can be taken to improve the current situation.

Using boundary critique can help to surface those perspectives and make them explicit in a way that opens them up for dialogue, leading to a greater understanding of ongoing tensions and conflicts and the potential to work collaboratively to create greater shared understanding. It can also be used effectively as part of first-person reflective practice to make sense of messy, confusing situations.

To demonstrate the use of boundary critique, we use an example from the English context where partnerships have been central to public health and health and social care since the 1990s (Dickinson and Glasby, 2010; Perkins et al., 2010). In 2010, the then new Coalition Conservative/Liberal Democrat UK government announced its intention to require local authorities to establish 'Health and Wellbeing Boards' as a governance structure for local level partnership working between local authorities, the NHS, and other stakeholders. These Boards, made law in April 2013 as a result of the Health and Social Care Act 2012, involve as a statutory minimum local authorities, NHS commissioners and consumer 'champions' known as Healthwatch.

One of the authors (Helen Wilding) works as coordinator and development lead for the wellbeing and health partnership arrangements in Newcastle upon Tyne in the North East of England. She has used critical systems heuristics (CSH) in combination with other tools from the soft systems tradition (including rich pictures, snappy systems, and simple PQR systems) to understand the multiple, co-existing perspectives on the current partnership and what it ought to be. Her reflective first-person action inquiry has been driven by the following questions:

- How do I understand and give expression to interrelationships of health partnerships (the 'is' world)?
- How do I appreciate and give expression to multiple perspectives that may inform a partnership system for health (the 'ought' world)?
- How do I give expression to my reflections on what facilitates (helps) and constrains (hinders) the realization of the 'ought' world of a partnership system for health?

Relevant work-based literature, informal 'real-world' observations, personal reflections, and conversations with stakeholders have been used as sources of data for the inquiry.

Understanding interrelationships (the 'is' world)

Interrelationships can be understood as constituting various levels, including the rich complicatedness of related entities as seen from any one perspective, the complexity of different perspectives, and conflicts among different perspectives.

Using 'snappy systems' as way of capturing perspectives

There are a variety of perspectives of the partnership that can be conveyed in terms of purposes. Some perspectives make a distinction between a broader partnership and the specific structure of the Health and Wellbeing Board, whereas others conflate the two. Some perspectives are not explicitly articulated but can be evident in the way people behave. Perspectives identified either locally or from a national perspective include the following quick-fire list of 'snappy systems':

The partnership is:

- A system to improve wellbeing and health.
- A system to cut costs.
- A system to coordinate the activities of individual organizations.
- A system to bring democratic accountability into the NHS at a local level.
- A system to ensure organizational or sectoral interests are protected.
- A system to legitimize and authorize a plan.
- A system to implement government requirements.
- A system to develop a multi-agency strategy.
- A system to create an effective and efficient health and care system.

The final snappy system listed above – a system to create an effective and efficient health and care system – is the perspective of local partnership arrangements that is most evident in government discourse and therefore has a powerful influence on many stakeholders and the business carried out by Health and Wellbeing Boards. It will therefore be used below to explore further the dynamics of the system.

Using PQR to express a system purpose definition ('simple systems')

As described earlier, PQR can be used to express the purpose (P) of a system more rigorously by making explicit why it is necessary (R) and how it might be achieved (Q).

In the case of this example, a simple system could be:

To create an effective and efficient health and care system **(P)**
by means of integrated commissioning and delivery **(Q)**
in order to ensure that citizens with, or at risk of, illness or disability have good outcomes **(R)**.

Using critical systems heuristics to further unfold purpose and conflict

Other dimensions of the system in operation, including the dimensions of potential conflict, can be explored through further unfolding using a critical systems heuristics (CSH) reference system. It is used below based on an understanding of how the PQR system is actually operating, with particular regards to measures of success (motivation), the decision environment in which it is operating (control), the types of expertise being drawn in (knowledge), and the underpinning rationale and political space in which it operates (legitimacy).

Sources of motivation

Most stakeholders agree that the beneficiaries of the system are people who are at high risk of, or who already have, illness or disability. The purpose is to reduce the risk of illness or minimize its progression and maximize the individual's quality of life. In practice, there is a particular focus on those with long-term conditions because these are the people who place the greatest demand on health and care services.

The beneficiaries of the system would understand success in terms of improvements to, or less deterioration in, their health and wellbeing. While this is a very individual focus, in practice the measures of improvement in use are at an aggregate level, such as a reduction in the number of hospital admissions. In the current UK context, measures of improvement are not only about the outcomes for the individuals but also about how much the intervention costs (or saves) the health and care system. At a time when budgets are getting tighter, these surrogate measures placing an emphasis on the financial viability of the health and care system. or the organizations within it, can override the focus on outcomes for individuals.

Sources of control

Most of those involved in discussing who make decisions would espouse the view that patients or service users themselves should be integral to decision-making. However, in practice, it is recognized that the Health and Wellbeing Board will make the decisions and in doing so talk about the importance of accountability to patients or service users. Some members of local Health and Wellbeing Boards are elected local politicians and are therefore held accountable via the mechanisms of representative democracy. Others are accountable through their professional standards and by adhering to principles of integrity in public life.

However, the degree of control available to Health and Wellbeing Boards is set in a context of decisions made at a national level, the most prominent of which are decisions about financial resources. Other elements of the decision environment influence decision-making in more subtle ways. For example, national requirements stipulate what plans must be produced and approved and may also dictate timescales and even format/structure. The strong influence of this decision environment can distract a local level Health and Wellbeing Board from developing its own sense of purpose and way of working.

Sources of knowledge

In order to make decisions, a Health and Wellbeing Board depends on knowledge support from a number of different sources of knowledge. A key source of knowledge is an evidence base of local need. The legislation requires local Health and Wellbeing Boards to produce a 'Joint Strategic Needs Assessment' that draws together a variety of data sources to inform the understanding of the local population and their needs.

Additional sources of knowledge are available from within the statutory membership of the Health and Wellbeing Board itself, for example:

* Elected members provide knowledge on what is important to local people.
* Healthwatch provides knowledge on what is important to patients and services users.
* The Director of Public Health provides research evidence on what it is important to prioritize and what works.
* The Clinical Chair of the NHS Clinical Commissioning Group provides insights gained from being part of the GP practice community.

The Board will also draw on the expertise of officers and managers within the partner organizations for knowledge such as policy advice, planning processes, and change management.

Each of these different sources of 'knowledge' provides different guarantees of validity, including, for example, knowledge drawing on experience, theory-informed knowledge, and knowledge formalized by scientific research. At times, there can be conflict between the different types of knowledge along a number of different dimensions:

- Objective and scientific or subjective and experiential.
- Reductionist (focusing on parts) or holistic (focusing on wholes).
- Linear models of causation or emergent models of causation.
- Dogmatism (favouring a dominant perspective) or appreciating multiple perspectives.
- Pathogenic focus (what reduces disease) or salutogenic focus (what creates health).

Sources of legitimacy

Some stakeholders who can be affected by the partnership's work are not included in the nationally defined statutory members of Health and Wellbeing Boards. Examples include voluntary and community sector organizations and NHS Foundation Trusts. Some local areas, including Newcastle upon Tyne, have responded to this issue by inviting these partners and others to join the Health and Wellbeing Board itself, thus expanding the range of people involved in decision-making and the range of knowledge and perspectives that can be drawn on in discussions and decision-making.

However, given there is always a limit to the number of people who can operate effectively together as a Board, some will remain witness to the work of the partnership. Wider stakeholders, each with their own perspectives of the purpose, use a variety of ways to raise awareness of their interests, which may include letters to the Chair, recommending particular content for the Joint Strategic Needs Assessment, and highlighting issues that should be included in a local strategy.

The worldview that possibly underpins this system is the 'negative conception' of health as lack of illness (Holland, 2007) rather than the positive conception of health as wellbeing. Such a rationale draws predominantly on the biomedical model of health but also engages to some degree with a social-behavioural model. It therefore prioritizes the intervention of services, whether for treatment, care or behaviour change at an individual level. This is a manifestation of the institutionalization of the biomedical model of health in the UK (Smith, 2013) and the tendency for lifestyle drift (Popay et al., 2010).

In associating the 'health system' with health and care systems, the possibilities for public health action are reduced (Hunter and Perkins, 2014). It limits the possibilities for improvements in wellbeing and health that can be offered by drawing on a more ecological model of public health (Rayner and Lang, 2012), such as those that intervene at population or community level. It marginalizes other worldviews based on a more salutogenic perspective of health as wellbeing. It also limits opportunities for reducing the health gap that persists across the social gradient. It is this worldview that is evident in the discourse that gives expression to the way Newcastle upon Tyne would like to work – the way some stakeholders, including myself, think it *ought* to be.

Engaging with multiple perspectives (the 'ought' world)

Newcastle upon Tyne in the UK had already formed a Wellbeing and Health Partnership prior to the statutory requirement to introduce Health and Wellbeing Boards. However, it saw the introduction of the new statutory requirement as an opportunity for a major review not just of partnership structures but how partnership working is conducted in the city. As a long-standing designated city within the World Health Organization's European Healthy Cities Network, the council also saw the opportunity to embed learning on inter-sectoral governance for wellbeing and health (Kickbusch and Gleicher, 2012). Perhaps the most obvious manifestation of this position was the city's choice to refer to its statutory 'health and wellbeing board' as the Wellbeing for Life Board in order to emphasize a positive conception of health and the importance of a life course approach.

Drawing on local documentation (such as Cabinet reports, Policy Cabinet papers, the local Wellbeing for Life Strategy) and conversations with colleagues as part of my experience of being part of partnership working in Newcastle upon Tyne, it is possible to use both PQR and critical systems heuristics to elaborate a normative ('ought') model for the partnership and the role of the Wellbeing for Life Board as a key part of it.

The purpose of the 'partnership system' can be phrased (using PQR) as follows:

To create the conditions for positive wellbeing and good health for all (P)
by means of a comprehensive health in all policies approach to urban development (Q)
in order to ensure our current and future citizens flourish (R)

This perspective positions the boundaries of the system as the geography of the local authority area, and consistent with a healthy city approach based on an ecological model of health incorporates the range of actions that can be taken by different sectors, institutions, politicians, and different professionals. It has a population focus and more of a salutogenic underpinning.

This distinction between narrower and wider perspectives on health and action has been made elsewhere. For example, Hunter (2007) distinguishes managing *for* health from 'health management'. Similarly, the WHO has distinguished governance *for* health and wellbeing from 'health governance' (Kickbusch and Gleicher, 2012). The distinction highlighted here could therefore be that of an emphasis on partnership *for* wellbeing and health, rather than a 'health partnership' – in short, distinguishing systems *for* health, as against 'health systems'.

As previously, critical system heuristics can be used to further unfold this perspective.

Sources of motivation

The beneficiaries of the partnership ought to be the current and future citizens of the city. By creating conditions for positive wellbeing and good health and focusing on protective factors, it is anticipated that citizens will have a good quality of life and will be less likely to develop disease or illness. Measures of improvement ought to be concerned with those that reflect citizens' experiences of the conditions of everyday life as well as more traditional health measures such as life expectancy and disease prevalence or incidence. Importantly, the focus on equity means that it is not just improvements in averages that ought to be important but signs that gaps are narrowing.

Sources of control

Consistent with an emphasis on participatory democracy, local citizens ought to be in control of improving their conditions of everyday life through active participation in broader policy processes, service improvement, and more localized neighbourhood change. This leads to a different normative perspective of the role of the Wellbeing for Life Board, which needs to focus on changing institutions and ways of working that build citizen participation and activism so that they can create *their* healthy city.

In order for this change to happen, fewer constraints should be imposed through the decision environment by national government. There ought to be additional resources devolved to local control and more freedoms and flexibilities to respond to issues that the local population think are important.

Sources of knowledge

The statutory requirement to produce a Joint Strategic Needs Assessment ought to be implemented in a way that supports a comprehensive health-in-all-policies approach to urban development and encourages active participation in the processes of interpreting and making judgements about the relative importance of different issues. Knowledge ought to be co-produced through dialogue, bringing together people from organizations and local people to learn with and from each other.

There ought to be a greater focus on knowledge of *how* to work to develop and nurture community participation, rather than only defining problems and their solutions through data collected and analysed by experts.

All those involved in providing knowledge ought to avoid:

- Reductionism: selecting only factors that can easily be worked with.
- Dogmatism: uncritically adopting a perspective that conforms with dominant values.
- Populism: allowing the loudest collective voice to be sole guarantor.
- Scientism: sole reliance on objective and statistical 'fact' from conventional evidence-based data derived from, for example, randomized control trials.
- Favouring pathogenic factors over salutogenic ones.

Sources of legitimacy

It is important to note that this normative *ought* perspective – to create the conditions for positive wellbeing and good health for all – does not negate the need to create an effective and efficient health and care system. Such a system is a necessary (but not sufficient) condition for positive wellbeing and good health for all. The normative perspective respects and values the biophysical model of health but also expands beyond it to value other dimensions in an integrated ecological approach (Rayner and Lang, 2012), including the value of salutogenic factors (Kickbusch, 1996).

People advocating and lobbying on behalf of those who cannot participate directly themselves ought to be able to participate in making change happen, rather than just remain as witnesses. In particular, Healthwatch and Local Government Overview and Scrutiny arrangements play a key role in this.

A key political tension to manage is the intent to be inclusive while remaining resolutely innovative, which in practice can be very difficult.

Reflecting on boundary judgements (contrasting 'is' with 'ought')

Given that the 'is' world is a process of becoming rather than a static position, it is important to consider whether actions are taking us near to, or distracting us from, the 'ought' ideal model. It is important to consider what helps (opportunities) and what hinders (challenges) the implementation of the ideal system.

Sources of motivation

'Know Your City', a key online document (www.knownewcastle.org.uk) that provides a profile of the people in Newcastle upon Tyne as well as the factors that affect their quality of life, includes continually updated data from 2011 that allows for judgements to be made on whether or not conditions for health and wellbeing are improving or deteriorating. It prompts steps to incorporate information that goes beyond averages and to look at differences across the social gradient wherever data allow. While these measures of improvement are available to people both to reflect on past performance and guide future action, there is little evidence that they are being used in the way originally intended as a resource for those involved in urban development. It is therefore important not just to invest in the production of the 'Know Your City' document itself but to invest in policy practice changes that draw on its use.

Sources of control

Since it was established, the Newcastle Wellbeing for Life Board has endeavoured to take actions that move it towards its normative ideal. However, this has been difficult due to the hegemony of the national government perspective affecting the decision environment, along with austerity impacting on the resources available to facilitate the level of systemic change required.

Newcastle is working with neighbouring authorities to establish a devolution agreement with government, as this would offer the potential for greater localized decision-making and freedom to act more closely with local people. This work is informed by the North East Commission for Health and Social Care, which reported in late 2016 (NECA, 2016). This work presents opportunities to reshape the decision environment.

Sources of expertise

The normative system has a very pluralistic view of the knowledge needed, who 'produces' it and who 'uses' it. Within Newcastle, there have been some good examples of participatory approaches to knowledge production both in research-led and practice-led initiatives. However, these have been small-scale and fragmented and have not yet resulted in a significant shift away from a paradigm that favours knowledge held by those with power or authority.

Sources of legitimacy

There is a range of cultural challenges in terms of moving towards the 'ideal' system. This involves being more open, accepting of fallibility and uncertainty, and people in organizations being willing and able to learn with and from local people through continuity of dialogue. In Newcastle, there have been some small-scale *ad hoc* initiatives to make that change but this has not yet been sustained long enough to create the required culture change.

Summary of case study

The case study reported on here is a reflective first-person action inquiry on the part of one of the co-authors, Helen Wilding. The three dimensions of the STiP framework (informed by boundary critique) provide a helpful reflective tool for informing not just this largely individual inquiry, but future engagement and praxis with colleagues on the issue of generating systems for health partnership. Thus, a further iteration of the three (first-person) research questions (outlined above on pp. 46–54) can be expressed in terms of a wider (collective and concerted) purposeful systemic inquiry:

- What are the key interrelationships regarding health partnerships (the 'is' world)?
- How might multiple perspectives be engaged towards formulating an improved model of partnership (the 'ought' world)?
- What facilitates (helps) and constrains (hinders) the realization of a system for health partnerships?

Moreover, such questions, while helpful at different scales of purposeful inquiry in a specific UK-based context, are also helpful and appropriate across international scales (associated with partnerships associated with WHO) and variable contexts, including within low-income countries.

Notes

1 In his 2006 paper, Midgley acknowledges Werner Ulrich as the originator of the term 'boundary critique' (cf. Ulrich, 1996a). Midgley goes on to theorize his own understanding of boundary critique in terms of how value judgements associated with issues and stakeholders (for example, in the public health sector) can be marginalized through ritualized practices marking a distinction that he makes between values that are 'sacred' and values that are 'profane' (Midgley, 1992, 2006).
2 Many of the OU STiP mature-age part-time students (cf. Open University, 2012) come from health-related professions; evidence from dissertations and subsequent online discussions on the STiP alumni LinkedIn site suggest an enthused uptake of CSH ideas.

References

Armson, R. (2011) *Growing Wings on the Way: Systems thinking for messy situations*. Axminster: Triarchy Press.

Blackmore, C., Ison, R. and Reynolds, M. (2014) Thinking differently about sustainability: experiences from the UK Open University, in W.L. Filho, U. Azeiteiro, F. Alves and S. Caeiro (eds.) *Integrating Sustainability Thinking in Science and Engineering Curricula* (pp. 613–30). Dordrecht: Springer [http://oro.open.ac.uk/40867/].

Checkland, P. (1981) *Systems Thinking, Systems Practice*. London: Wiley.

Checkland, P. (2011) Autobiographical retrospectives: learning your way to 'action to improve' – the development of soft systems thinking and soft systems methodology, *International Journal of General Systems*, 40 (5): 487–512.

Checkland, P. and Scholes, J. (1990) *Soft Systems Methodology in Action*. London: Wiley.

Churchman, C.W. (1970) Operations research as a profession, *Management Science*, 17 (2): B37–B53.

Churchman, C.W. (1979) *The Systems Approach and Its Enemies*. New York: Basic Books.

Cook, S.N. and Wagenaar, H. (2012) Navigating the eternally unfolding present toward an epistemology of practice, *American Review of Public Administration*, 42 (1): 3–38.

Dickinson, H. and Glasby, J. (2010) Why partnership working doesn't work: pitfalls, problems and possibilities in English health and social care, *Public Management Review*, 12 (9): 811–28.

Guyatt, G. and Rennie, D. (eds.) (2002) *Users' Guides to the Medical Literature: A manual for evidence-based clinical practice*. Chicago, IL: AMA Press.

Holland, S. (2007) *Public Health Ethics*. Cambridge: Polity Press.

Hunter, D.J. (2007) Exploring managing for health, in D.J. Hunter (ed.) *Managing for Health* (pp. 54–79). Abingdon: Routledge.

Hunter, D.J. and Perkins, N. (2014) *Partnership Working in Public Health*. Bristol: Policy Press.

Kickbusch, I. (1996) Tribute to Aaron Antonovsky: 'What creates health', *Health Promotion International*, 11 (1): 5–6.

Kickbusch, I. and Gleicher, D. (2012) *Governance for Health in the 21st Century*. Copenhagen: WHO Regional Office for Europe [available at: http://www.euro.who.int/en/what-we-publish/abstracts/governance-for-health-in-the-21st-century; accessed 22 January 2013].

Lang, T. and Rayner, G. (2012) Ecological public health: the 21st century's big idea?, *British Medical Journal*, 345: e5466 [doi: 10.1136/bmj.e5466.].

Lowndes, V. and Skelcher, C. (1998) The dynamics of multi-organisational partnerships: an analysis of changing modes of governance, *Public Administration*, 76 (2): 313–33.

Melnyk, B.M. and Fineout-Overholt, E. (eds.) (2011) *Evidence-based Practice in Nursing and Healthcare: A guide to best practice*, 2nd edn. Philadelphia, PA: Lippincott Williams & Wilkins.

Midgley, G. (1992) The sacred and the profane in critical systems thinking, *Systems Practice*, 5 (1): 5–16.

Midgley, G. (2006) Systemic intervention for public health, *American Journal of Public Health*, 96 (3): 466–72.

NECA (North East Combined Authority) (2016) *Commission for Health and Social Care Integration in the North East* [available at: http://www.northeastca.gov.uk/devolution/commission-health-and-social-care-integration-north-east; accessed April 2017].

Nuyens, Y. (2007) Setting priorities for health research: lessons from low- and middle-income countries, *Bulletin of the World Health Organization*, 85 (4): 319–21.

Open University (2012) *Thinking Strategically: Systems tools for managing change*. Study Guide (2nd edn.). 30 credit Open University module (code TU811) for the postgraduate Systems Thinking in Practice programme. Milton Keynes: The Open University [available at: http://www.open.ac.uk/postgraduate/modules/tu811; accessed 14 July 2017].

Perkins, N., Smith, K.E., Hunter, D.J., Bambra, C. and Joyce, K. (2010) 'What counts is what works'? New Labour and partnerships in public health, *Policy and Politics*, 38 (1): 101–17.

Popay, J., Whitehead, M. and Hunter, D.J. (2010) Injustice is killing people on a large scale – but what is to be done about it?, *Journal of Public Health*, 32 (2): 148–9.

Porter, M.E. (2010) What is value in health care?, *New England Journal of Medicine*, 363 (26): 2477–81.

Rayner, G. and Lang, T. (2012) *Ecological Public Health: Reshaping the conditions for good health*. Abingdon: Routledge.

Real, KM (2015) Retrieved from: http://realkm.com/2015/11/12/knowledge-brokers-connecting-research-and-practice/ [accessed December 2015].

Reynolds, M. (2007) Evaluation based on critical systems heuristics, in B. Williams and I. Imam (eds.) *Using Systems Concepts in Evaluation: An expert anthology* (pp. 101–22). Point Reyes, CA: Edge Press.

Reynolds, M. (2008) Getting a grip: a critical systems framework for corporate responsibility, *Systems Research and Behavioural Science*, 25 (3): 383–95.

Reynolds, M. (2014a) Triple-loop learning and conversing with reality, *Kybernetes*, 43 (9/10): 1381–91 [http://oro.open.ac.uk/41373/].

Reynolds, M. (2014b) Equity-focused developmental evaluation using critical systems thinking, *Evaluation*, 20 (1): 75–95.

Smith, K. (2013) Institutional filters: the translation and re-circulation of ideas about health inequalities within policy, *Policy and Politics*, 41 (1): 81–100.

Ulrich, W. (1983) *Critical Heuristics of Social Planning: A new approach to practical philosophy*. Chichester: Wiley.

Ulrich, W. (1996a) *Critical Systems Thinking for Citizens: A research proposal*. CSS Research Memorandum #10. Hull: Centre for Systems Studies.

Ulrich, W. (1996b) *A Primer to Critical Systems Heuristics for Action Researchers*. Hull: Centre for Systems Studies.

Ulrich, W. (1998) *Systems Thinking as if People Mattered: Critical systems thinking for citizens and managers*. Working Paper #23, Lincoln School of Management, University of Lincolnshire and Humberside.

Ulrich, W. (2000) Reflective practice in the civil society: the contribution of critically systemic thinking, *Reflective Practice*, 1 (2): 247–68.

Ulrich, W. (2001) The quest for competence in systemic research and practice, *Systems Research and Behavioural Science*, 18 (1): 3–28.

Ulrich, W. (2002) Boundary critique, in H.G. Daellenbach and R.L. Flood (eds.) *The Informed Student Guide to Management Science* (pp. 41–2). London: Thomson Learning.

Ulrich, W. (2003) Beyond methodology choice: critical systems thinking as critically systemic discourse, *Journal of the Operational Research Society*, 54 (4): 325–42.

Ulrich, W. and Reynolds, M. (2010) Critical systems heuristics, in M. Reynolds and S. Holwell (eds.) *Systems Approaches to Managing Change: A practical guide* (pp. 243–92). London: Springer.

van Olmen, J., Marchal, B., Van Damme, W., Kegels, G. and Hill, P.S. (2012) Health systems frameworks in their political context: framing divergent agendas, *BMC Public Health*, 12: 774 [doi: 10.1186/1471-2458-12-774].

Walshe, K. and Rundall, T.G. (2001) Evidence-based management: from theory to practice in health care, *Milbank Quarterly*, 79 (3): 429–57.

World Health Organization (WHO) (2008) *The Tallin Charter: Health systems for health and wealth*. Copenhagen: WHO Regional Office for Europe [available at: http://www.euro.who.int/en/media-centre/events/events/2008/06/who-european-ministerial-conference-on-health-systems/documentation/conference-documents/the-tallinn-charter-health-systems-for-health-and-wealth; accessed 26 February 2016].

4 Soft systems methodology: an approach for stakeholder and researcher reflection on a problem

Kathy Kotiadis

Introduction

Many problematic situations are poorly understood as they emerge, with many stakeholders reaching completely different opinions about their cause and resolution. Often this is because the stakeholders involved in the situation are focused on their part of the system. In such circumstances, if the process of enquiry towards reaching action to alleviate the situation is not handled with all stakeholders in mind or their views consolidated, one might be left in a worse situation. Indeed, such ground is fertile for blame, conflict, and even a breakdown in working relationships that can eventually exacerbate the situation. Perhaps more importantly the resulting action may not have the universal support for implementation or actually resolve the situation. Soft systems methodology (SSM) is an organized learning system, able to support the development of a shared view of the problematic situation and an agreement towards feasible and desirable action.

SSM was introduced in the 1970s by Peter Checkland (Checkland, 1972) of the University of Lancaster (UK) in recognition of the importance that should be placed on the process of enquiry. Over the first 20 years, Checkland regularly redeveloped SSM, with real practice in mind, culminating with its final version in the 1990s: the four main activities version of SSM (Checkland, 1999a, 1999b). This final version, described in this chapter, enabled greater flexibility in industries outside of the private sector such as the Civil Service and the UK National Health Service (NHS).

One of the main advantages of SSM, over other analytical approaches, is that it does not require any software, or for the analyst(s) or those taking part to have a technical background. Essentially, anyone can be easily trained to use SSM. Hence, this chapter will aim to provide the basic training for its immediate application.

What does soft systems methodology do?

SSM offers a process of enquiry even when there is ambiguity or even disagreement about the problematic situation. To start with, one might not have a specific objective in mind but a willingness by the stakeholders to participate in the process of enquiry. The person or persons leading the enquiry could be among the stakeholders or be an outsider(s). In this chapter, I refer to this individual or group as the analysts. The SSM process of enquiry is best driven by its methodology, as it provides a helpful structure. The four main activities of the SSM methodology are as follows:

1. Finding out about a problem situation, including the cultural and political situation.
2. Formulating some relevant purposeful activity models.
3. Debating the situation, using the models, seeking from that debate both:

   • changes that could improve the situation and are regarded as both desirable and (culturally) feasible, and
   • accommodations between conflicting interests, which will enable action-to-improve to be taken.

4. Taking action in the situation to bring about improvement.

The remaining sections of this chapter will explain the approach and the paper-based tools supporting it. Subsequently, we will explore the practice of SSM as well as practical considerations and tips for successful application. Finally, a SSM real-world case study will be presented.

Soft systems methodology and its tools

The following sections will focus on each activity representing a stage in the enquiry.

Stage 1: Finding out about a problem situation

This stage can be very daunting, as it can be difficult to establish if you have found out enough about the problem situation. As this stage is not quantifiable, it is best thought of as a process of getting everyone who is part of the situation and could be part of the solution thinking and talking. At this point, the analyst might try and determine who should be involved and why. Given that the process aims for an eventual agreement towards action, it is best to involve a wide range of stakeholders so that implementation of any change in practice is widely accepted. For example, if the problem situation involves a health clinic, it might be wise to consider inviting staff involved in managing the clinic and the staff treating the patients. It might also be worth including representatives of patients groups in the process.

The stakeholders can be involved with the analysts on a one-to-one basis or as a group. This will largely depend on factors such as the number in the stakeholders group, the likelihood of being able to meet at the same time (e.g. geographical constraints or time commitments), as well as other practicalities. For example, in healthcare settings, shift work may prevent medical staff from being able to meet concurrently. In these sorts of situations, it is worth liaising with stakeholders over practicalities to ensure the most effective approach is adopted for information gathering and dissemination. These practicalities will be further discussed in a subsequent section. Of course, not everything can be established from the start of a study, since in many cases, as the study progresses and the problem situation becomes clearer, different types of information and further stakeholders can be identified. When the initial stakeholders are identified, the analyst should consider the approach or tools for the enquiry. SSM offers the following two approaches to finding out about a problem situation: Rich Picture Building and Analysis One, Two, and Three.

Rich Picture Building

The first approach, Rich Picture Building (Checkland, 1999b), is about recoding a conversation in a picture format. There are no rules on who should draw the picture or how the picture is drawn or any guidance on representations. The picture itself can include text, arrows, graphs, drawings of people, buildings, machinery, etc. Examples of SSM's rich pictures can be found online with some sources included in the references (e.g. Hindle et al., 2015; Open Learn, n.d.; Pham, 2014). A rich picture can be drawn by the analyst with the help of a stakeholder, in a one-to-one meeting, or with the help of a group of stakeholders in a workshop. The most important aspect of drawing the picture is that everyone participating understands its meaning, but there is no need to aim for a sophisticated drawing. Indeed, what could be described as unsophisticated drawings or rich pictures have been used by the UK Department of Health in government documents (Department of Health, 1997). The advantage of this mode is that it can capture information that is disjointed and complex human relationships that might be difficult to articulate in words.

Drawing rich pictures offers a starting point and often a blank canvas ensures that the participants appreciate the analyst is expecting them to portray their understanding. For example, it may be the case that no one knows at this point what they are looking for, but might have a vague idea of the symptoms of the emerging situation that requires a remedy. In such a case, discussing and recording these symptoms offers a sensible starting point. However, if there is a more precise starting point in mind, a preliminary rich picture can be brought in to get the stakeholders to focus on a particular aspect of the situation. For example, it might be a case of improving or evaluating some known aspect of the practice of the organization. Hence, the starting point of the problem situation might take many forms, which is why there are no rules to the design of rich pictures but it is expected that the completed picture should capture the situation, its main stakeholders and issues.

Analyses One, Two, and Three

These analyses can be undertaken instead of rich pictures or in addition to them. They can be thought of as a line of enquiry that can involve observations, interviews, workshops or other means of engaging with stakeholders to gain information. Analyses One, Two, and Three are otherwise respectively known as role analysis, social system analysis, and political system analysis.

Role analysis, or Analysis One, involves exploring three main roles: the role of the client (who has caused the study to take place), the role of the 'would be problem-solver' (who wants to do something about the situation), and the role of the problem-owner. All or some of these roles may overlap. Determining who the problem-owners are is not as simple as one might imagine in situations where there is ambiguity over what part of the system the problem situation belongs to. Indeed, the symptoms of a problem often 'travel' within systems, so this can become a case of finding the source and those individuals that can make decisions to improve that part of the system (owners).

Social system analysis, or Analysis Two, is based on the notion that a social system is a continually changing interaction of three elements: roles, norms, and values. Roles are social positions of importance to the problem situation that are institutionally defined or behaviourally defined. For example, in many surgical departments consultant surgeons were considered to have the highest status in the past. In recent years, however, other roles have emerged as equally important, such as that of hospital manager. Furthermore, a role is characterized by expected behaviour otherwise known as norms. This could be as simple as identifying the hierarchy and associated behaviours. For example, in the UK National Health Service, consultant surgeons lead a firm, which is a more junior group of surgeons. However, these firms change from year to year as surgeons move to new training posts. Surgeons value the experience gained by training with multiple surgeons. This would be the sort of information that would need to be captured through this analysis, if the problematic situation involved surgeons.

Political system analysis, or Analysis Three, is about understanding how power is expressed in a particular problematic situation. For example, power may be in the form of personal charisma or even membership of a particular committee. Understanding the roles within a problem situation, the typical behaviour of the stakeholders, and the allocation of power means that the analyst is better able to manage the stakeholders during the process and arrive at action that is agreeable to all, desirable, and feasible.

All three analyses complement each other and do not necessarily need to be undertaken in any particular order. The importance and time allocated to each analysis will depend on the problem situation. Also, the nature of some of the questions being asked should involve a certain amount of sensitivity, which could mean that some or all of this analysis may need to be done covertly (Pidd, 2007). A possible solution to this is taking advantage of the rich picture drawing session, if this is used, to observe stakeholders and ask leading questions as part of Analyses One, Two, and Three. This covert analysis is feasible, as stakeholders can be requested to feature within the drawings. During the role, social, and political system analyses, the analyst will become aware of organizational issues or tensions that might affect the development of the SSM model and ultimately the implementation of any action to alleviate the problem situation. This analysis is particularly important when the analyst is not familiar with the environment of interest.

Stage 2: Formulating some relevant purposeful activity models

The aim of Stage 2 is to engage with the stakeholders in a systematic way to produce a model that will be used for debate. The analyst should encourage debate throughout the entire process. Indeed, the aim should be to provide debate with a strategy for divergence and then convergence. This means that the analyst should be encouraging multiple views to be put forward and through debate, to narrow down these views to a single view that fits the stakeholder group. Obviously, this is easier and quicker to achieve in a workshop setting. Hence, from here on, the process will be described as if it takes place in a workshop setting. Alternative approaches to practice will also be discussed later in this chapter.

The model takes the form of a linked text diagram and is called a purposeful activity model (PAM) but can also be found in SSM literature as a conceptual model or a human activity model. The analyst will guide the stakeholders in using a set of tools aiming to generate the PAM. The first tool used is CATWOE – Customer(s), Actor(s), Transformation process, Worldview, Owner(s), Environmental constraints (Checkland, 1999b) – and involves defining each of these aspects but not in the order of the acronym.

CATWOE and root definition

The analyst should encourage the stakeholders participating to consider a number of transformations of inputs to outputs that could be thought of as directions for improvements. The question 'what area/aspect do you want to alter or improve?' may help participants make a start. This definition is the easiest to confuse with the transformation supported by a particular system, as most systems produce outputs. For example, the output of an outpatient clinic might be to diagnose and treat patients but the problematic situation requiring improvement might only relate to issues around the clinical knowledge of staff. In such a case, the transformation that one might record is the need for more extensive educational support for the outpatient clinic staff. The input here is educational support transformed into more extensive educational support for all clinical staff. A group might come up with different transformations that should all be debated and could all be explored individually if time permitted, but ideally these should be narrowed down to as few as possible and ultimately just one. Open or secret voting can be used to narrow these down. For each transformation, it is also important to consider the activities that might support it in very general terms at this stage and to answer the questions: How will it be done, and by what means? For example, how will we provide more extensive educational support? The answer to this could be by teaching/training and mentoring activities. This answer should be recorded alongside the Transformation (T).

Once the transformation has been determined through debate, it is sensible to consider what worldview makes the transformation process meaningful in the particular context. For example, why is it important to provide extensive educational support for all staff in this outpatient clinic? The answer to this question is the worldview. Here, the Worldview (W) might be that if all staff were trained better, the clinic would be able to offer immediate diagnoses and treatments to more patients.

All the subsequent definitions from the CATWOE acronym will fit around the T and W. The Customers are beneficiaries (or victims) of the T. The beneficiaries of more extensive educational support will be the staff and the patients. The Actors are those who will undertake the T and therefore activities that increase the educational support. The Actors could be the staff themselves, online (internet) training support and courses, clinical staff belonging to other clinics providing training, etc. The Owners are those who can stop the T and in this example could be the management of the clinic or a leading clinician. Finally, the Environmental constraints are the training budget, the availability of trainers, and so on.

The CATWOE definitions can then be put into a format that resembles a mission statement and represents the interpretation of the problem situation and its various facets. This mission statement is called a 'root definition'. One easy way to assemble a full root definition is to use the following template, which will need some adjustments to fit each situation:

> A {insert the owners here} owned system undertaken by {insert Actors here}, to support {insert Transformation here} by {insert How to do it activities} in order to {Insert Worldview} for the benefit of {insert Customers}, while recognizing the constraints of {insert Environmental constraints here}.

For example:

> A clinician-led training programme for the outpatient clinic undertaken by a combination of trainers and online courses, **to support a more extensive educational development programme *by* teaching and mentoring activities *in order for* the clinical staff to be able to offer immediate diagnoses and treatments to more patients** for the benefit of the staff and their patients, while recognizing the training budgetary constraints and the available time for training.

Alternatively, simply list the acronym with its definitions and the essence of the root definition, which is: 'Do T by "How to do it activities" in order to achieve W'. This core part of the root definition is seen in bold in the above example.

The purposeful activity model

Building the PAM brings to life the root definition. The analyst can encourage the stakeholders to be active in the development of the model. The model-building process involves brainstorming the activities that would have to be there to support the T (Figure 4.1). The activities in these models do not need to represent

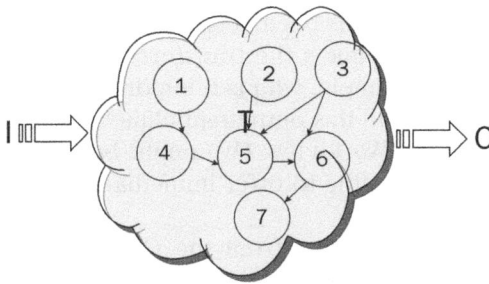

Figure 4.1 A conceptual view of the purposeful activity model (PAM) that describes the activities (shown as numbered circles) to support the transformation of inputs to outputs. I = inputs, T = Transformation, O = Outputs

only what already exists but stakeholders should be encouraged to think about what could be there, as activities to support the T. Hence, the development of the model can support exploratory thinking and discussion! The activities themselves are best recorded using a few words with a verb always being the first word, ideally in the imperative, which means that it will tell you what to do. For example, the activity 'assess staff competency' is much clearer than simply recording 'staff competency'. The number of activities to be recorded in each model will vary but a helpful guide is to aim for around seven so that the diagram is not too overcrowded but provides a detailed enough account of the activities supporting the T.

When the activities have emerged as a result of discussion, it is then sensible to structure these according to logical dependencies. This means you will need to consider their order and use a one-directional arrow to record this relationship (Figure 4.1). For example, the activity 'determine assessment for staff competency' will come before 'assess staff competency' (Figure 4.2). A single activity can be linked to many other activities using unidirectional arrows. The links should also be discussed as much as the activities. At the end, each activity should be numbered so these are more easily discussed and referred to accurately. As these diagrams are read from top (earlier activities) to bottom (dependent activities), the numbering should loosely follow that sequence (see Figures 4.1 and 4.2). In practice, these diagrams are quite messy to start with, so be prepared to redraw them and tidy them up at the end of the session!

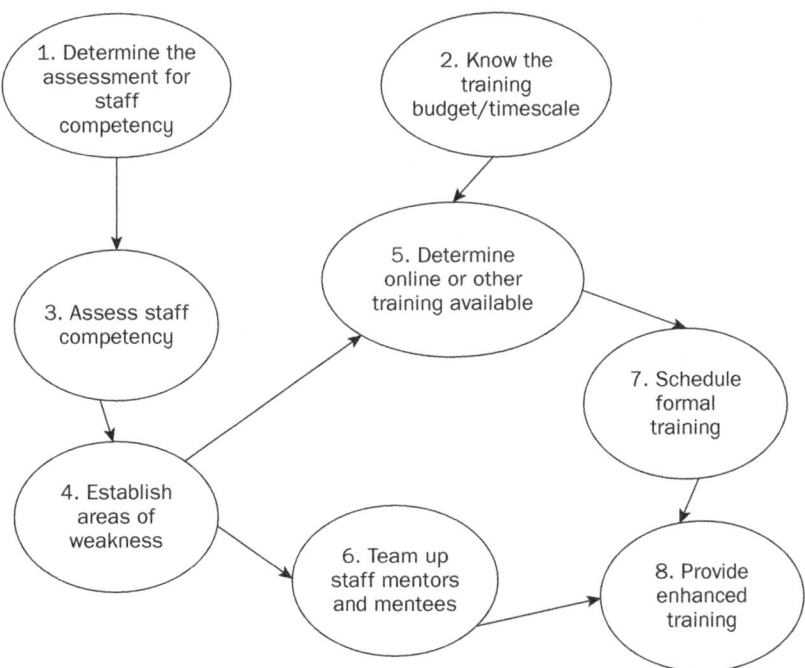

Figure 4.2 Example of a purposeful activity model (PAM)

Stages 3 and 4: Debating the situation and taking action

The purposeful activity model is particularly useful to support comparisons with the actual situation and help the stakeholders determine action that is desirable and feasible. The questions that can be put to stakeholders are: How is this currently done? How could it be done? Can you anticipate any reasons why this might not be accepted or work? Is there any conflict of interest? Can this conflict be resolved? By the end of the debate, the stakeholders should have recorded a list of actions to take forward that are considered desirable and feasible. Ideally, the action recorded should be linked to a named person or persons responsible as well as timelines for implementation. However, that is not the end of the debate or the recording of action, as more can emerge through the evaluation of the PAM, described next.

Evaluating the performance of the purposeful activity model

Once the PAM is debated and provisionally agreed as a proposed system, albeit directed towards taking action, we must also consider its evaluation. This will enable tracking its success or failure and provide the stakeholders with an opportunity to intervene as early as possible should the action not result in the desired effect. The evaluation of the PAM can be recorded as a system, which we will refer to as the performance measurement model (PMM) (Kotiadis et al., 2013). The PMM is a more recent extension to SSM, used to judge the performance of the PAM. Designing this system can also be an activity undertaken by the stakeholder group, and it enables another opportunity for debate and to ensure commitment to taking forward the action that has emerged from debating the PAM.

The starting point is to define the measures of performance, otherwise known collectively as the five E's:

- *Efficacy*: Does the means work? Or checking the output is produced.
- *Efficiency*: Amount of output divided by amount of resources needed. Or checking whether minimum resources are used to obtain it.
- *Effectiveness*: Is the Transformation (T) meeting the longer-term aim (i.e. the worldview (W))?
- *Ethicality*: Is the Transformation (T) morally correct?
- *Elegance*: Is this an aesthetically pleasing Transformation (T)? These days this criterion could also link T to sustainability.

In practice, most SSM studies report the use of the first three measures of performance with the latter two not fitting many situations. Additionally, these can be defined at a much earlier stage (i.e. during CATWOE), provided they are considered again after the development of the PAM. Taking forward our example, we could define the measures of performances as follows:

- *Efficacy*: Are the staff receiving enhanced training?
- *Efficiency*: Is the training/development undertaken within the budget and making use of freely available resources in the first instance?
- *Effectiveness*: Is the enhanced educational development programme enabling the clinical staff to offer immediate diagnoses and treatments to more patients?

- *Ethicality*: Is a more extensive educational development programme morally correct? There could be many arguments here about it being morally correct, as equipping the staff with knowledge enables them to provide a better service and to make fewer mistakes. However, other aspects might be worth considering such as whether staff will be paid for undertaking training or not.
- *Elegance*: Is this extensive educational development programme an aesthetically pleasing Transformation (T)? One could argue here that as more staff become better trained, they might be more efficient in their use of resources (e.g. instruments) or utilize shorter and more effective appointment slots making the view of the system more aesthetically pleasing. In any case, it would be difficult to argue that any aspect of this example of T is aesthetically displeasing!

Having defined the performance criteria or E's, the stakeholders can begin to construct the PMM. The approach to constructing it is very similar to the construction of the PAM. With the E's in mind, the stakeholders can reflect on how each activity, in the PAM, can be evaluated. Initially, they can be asked to consider specific monitoring activities, to observe and record information, which might lead to an evaluation. Where possible, these activities should begin with 'monitor' and then list the element to be monitored. For example, for our PAM Activity 3 (assess staff competency), one might *monitor the number of staff assessed*. Stakeholders can then be asked to consider what they would determine based on each of the monitoring activities or their combinations. This should be recorded in the format 'determine if'. Following on from our monitoring activity, we might record *determine if more staff can be assessed* (Figure 4.3). Where possible, the monitoring activities should be listed first and these should be linked according to logical dependencies to the 'determine if' activities. Similar to the core PAM, each activity in the PMM is circled and assigned a letter of the alphabet (rather than a number used in the core PAM). The generic relationship between the PAM and the PMM can be seen in Figure 4.4.

The construction of the PMM will also result in debate about the problematic situation but from the angle of what evaluation activities should be in place to know if it has been resolved. This in itself is very likely to require action, especially if new activities are introduced in the PAM. For example, it might be a case of setting up monitoring activities, targets and an evaluation process. Once again, to determine this action the stakeholders can be asked to identify what monitoring activities are in place currently and what action might be needed to expand these if necessary.

Each root definition, the activities of the corresponding PAM and its PMM result from a particular interaction. That means that there is no such thing as a correct or incorrect definition or model (PAM or PMM). Indeed, each SSM intervention is useful and holds true for that particular problematic situation and its stakeholders. You may have found yourselves disagreeing with the examples provided or believing they are incomplete. In the next section, we will explore the practice of SSM, especially with regard to practical considerations.

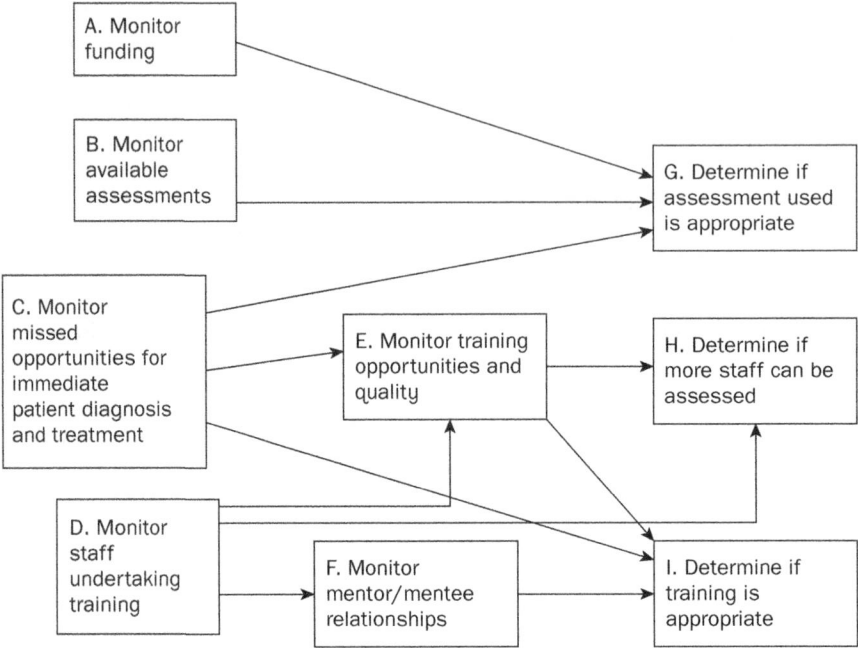

Figure 4.3 Example of a performance measurement model (PMM)

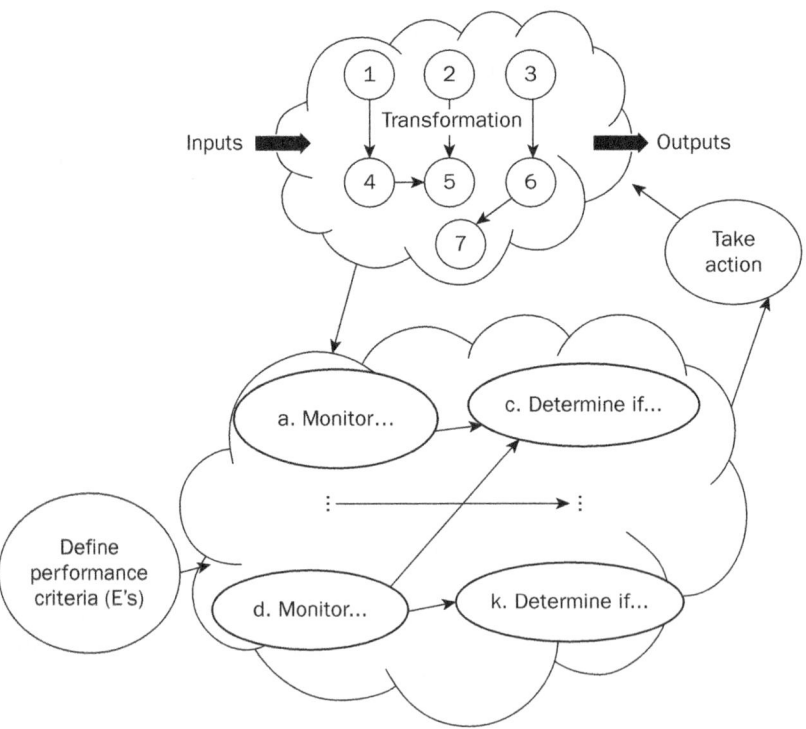

Figure 4.4 The relationship between the purposeful activity model (PAM) and the performance measurement model (PMM)

The practice of SSM

The practice of SSM should always be moulded to the situation of interest and the stakeholders involved. For example, at one end of the scale, an analyst may decide not to be explicit about the use of SSM or its tools and may use them subconsciously to inform his or her line of investigation and questioning. At the other end of the scale, another situation may be more suited to a 'Do it Yourself' (DIY) approach where a lead stakeholder may simply explain the process to the rest of the stakeholders and ask them to get on with it! Most reported practice falls somewhere in the middle with stakeholders participating under the guidance of an analyst or a group of analysts, taking on the role of the SSM facilitator. The following practical advice may be useful to those embarking on an SSM intervention:

Divide and conquer. Many investigations unearth new problematic situations that the stakeholders often want to resolve on top of the one undertaken. In such circumstances, it is worth liaising with stakeholders over the problematic situation to take forward. It might be a case of eliciting a new group of stakeholders to tackle the new situation or prioritizing with the same group the problem area to be tackled first.

Tailor to groups or individuals. There are clear advantages to involving stakeholders in SSM interventions concurrently. One of these advantages is the opportunity for collaboration and debate. Another advantage is arriving at common and agreed outputs. However, if for whatever reason that is not possible, the analyst can engage and undertake the process with individuals. If the same transformation is explored, the analyst may then try to combine the information into an agreeable model. Going back and forth between stakeholders to establish action may be challenging and time-consuming. However, there are advantages of exploring a problem with individual stakeholders. A very knowledgeable but shy stakeholder might provide more insights outside of a workshop. If stakeholders are known to argue or contradict each other when in the same room, it might be quicker to resolve the problematic situation by involving them on an individual level. One might even anonymize the contributions in such extreme cases!

Practise facilitation and soft skills. If a group of stakeholders is involved concurrently in what we would refer to as a workshop, the analyst will benefit from developing and practising facilitation skills. Engaging with a group concurrently requires an ability to maintain order and progress in a timely fashion. We have all experienced situations in which debate seems to go in circles or off topic. The analyst in such cases must have strategies to deal with this. For example, having a board to record issues that are not directly relevant for another time is one strategy. In addition, if a conversation is not resolving itself because information is not available at the time, the analyst might ask the stakeholders to consider all views and agree action for each possibility, until such a time when the information becomes available. The analyst facilitating the workshop should ensure that participants are able to contribute by encouraging multiple views and managing conflict so that it does not get out of hand. In my experience of running workshops, I have found that a clear workshop agenda with sensible timings for each activity helps stakeholders take responsibility for the outputs and focus on the aims of the workshop.

Practise SSM before engaging. Similar to the advice above, preparation in SSM goes a long way! SSM tools can appear simple but it's always sensible for the novice SSM analyst to consider the situation and attempt to construct the definitions and tools beforehand. These will probably never see the light of day but will enable the analyst to feel more secure during the intervention. This is particularly useful advice in the development of rich pictures, if they are used. For example, the analyst might develop some drawings for various issues or roles beforehand that he or she is aware of. Another practice that I have found useful is to bring a list of verbs to the workshop that can be used by the stakeholders to complete the description of each activity in the PAM.

A case study of soft systems methodology in health care

The case study described here is a real SSM intervention, undertaken in two parts, approximately three years apart, the first in 2006 and the second in 2009. This could be described as an unusual SSM study with deviations from described practice, even perhaps with earlier ones in this chapter, but that is the sort of artistic licence that one must take in one's stride in order to satisfy the needs of the customers with the resources at hand and the time available.

Although I was the SSM analyst leading the intervention, I was in the fortunate position of being supported by other analysts who did not have an SSM background at the time of the intervention. The first part in 2006 focused on the development of the PAM, while the second part in 2009 focused on the PMM and debate. The reason that the PMM was developed so much later than the PAM was because this was the first ever research proposing the PMM for SSM interventions. This case study was a turning point for SSM practice.

In 2006, a multidisciplinary colorectal cancer team of surgeons based at a hospital in the South of England wanted to explore their multidisciplinary team (MDT) function in light of an upcoming evaluation. Multidisciplinary medical teams are recognized as central decision points in the patient pathway, and the quality of information sharing in these meetings is an important factor in creating beneficiary outcomes for patients (Kane and Luz, 2011). For a more detailed explanation of the activities undertaken in MDT meetings, interested readers are referred to Kane and Luz (2011).

The aim of the study in 2006 was to determine if and what action should be taken to bring about improvements to the MDT meetings and their organization. The SSM intervention in 2006 involved observations of the MDT meetings, interviews with members of the team, and a facilitated workshop. In 2009, we approached the MDT again to find out if and what type of action had been taken since the first workshop and to explore whether a performance-monitoring element, that would take the form of a PMM, was desirable now that the system was more established. The 2009 part of the SSM intervention involved an observation of the MDT and a workshop.

The following sections provide an overview of the process and outputs of each intervention starting with the earliest one in 2006. However, this description is more of a summary demonstrating the tools and process. For a fuller account, interested readers are referred to Kotiadis et al. (2013).

Improving the multidisciplinary team meetings

Background information (2006)

The background information was obtained covertly through observation or informal interviews to understand the situation, the main issues, and stakeholders (Analyses One, Two, Three). In this case study, rich pictures were not developed. The colorectal MDT meetings were held weekly in the same venue with a permanent core membership consisting of the colorectal consultant surgeons (four in 2006), a histopathologist, an oncologist, a radiologist, nurse practitioners, MDT coordinators, and other junior doctors belonging to the colorectal firms (each consultant represents a firm). The seating was arranged in rows, with the most senior consultants at the front and other members depending on the 'pecking order' in subsequent rows. The flow of the conversation and decision-making was firmly at the front of the room, rather than including everyone in the room.

The MDT meeting operates as a pull system in which the coordinators have the responsibility of creating the patient list for each multidisciplinary meeting. For each patient, information such as the case notes and the results from all clinical investigations (for example, test results and imaging) are assembled. The patient selection process is largely dependent on the patients' time in the system and the availability of the results of the investigations. It is important that patients progress through the system in a timely fashion so the hospital meets waiting time targets set by the government.

In 2006, the consultant surgeons expressed some concerns about their MDT function and suggested areas of improvement, including the availability of patients' results and the equipment available in the room. Preliminary investigations revealed a lack of relevant documentation and a fragmented understanding of the MDT by its membership. It soon emerged that it was important to arrive at a common description of the MDT and its function, and so it was decided to organize a facilitated workshop for this task. SSM would be used to develop a shared understanding while accounting for multiple stakeholder involvement and diverse views. The following section provides a brief description of the workshop.

Workshop 1 (2006): The development of CATWOE and PAM

The analysts assembled the information extracted from observations of the MDT and interviews with the team members into preliminary SSM definitions (CATWOE, root definition) to be used in the workshop to aid the process of developing models of the MDT at operational level. The group as a whole had ownership of the MDT at the operational level and therefore provided more opportunities for subsequent action (implementation of change).

During the workshop, the SSM tool outputs (CATWOE, root definition and PAM) were debated and the necessary modifications were made to ensure that they were representative of the views of the MDT members regarding the optimal function and organization of their meeting (Table 4.1 shows the final definitions). The PAM constructed during the workshop was considered to be by all those attending as something to aim for (Figure 4.5).

Table 4.1 The CATWOE definition for the MDT meetings

CATWOE	Operational function definitions
Customers	Patients
Actors	MDT participants
Transformation process	The need for multidisciplinary and patient-centred diagnosis, treatment and care plan is met by presenting, evaluating, and deciding upon the results of diagnostic investigations
Worldview	A belief that multidisciplinary and patient-centred decision-making enables better diagnosis, treatment, and care plans that lead to improved patient outcomes
Owner	MDT participants, UK National Health Service Trust (Hospital), Department of Health
Environmental constraints	Clinical guidelines, physical and technological facilities available for the meeting, MDT participants' timetable, patient target times
The E's – performance measures	*Efficacy*: Does the MDT meeting provide multidisciplinary and patient-centred clinical decision-making (diagnosis, treatment, and a care plan) for each patient? *Efficiency*: Are the minimum resources used to obtain multidisciplinary and patient-centred clinical decision-making? *Effectiveness*: Does multidisciplinary and patient-centred clinical decision-making lead to improved patient outcomes?

Source: Kotiadis et al. (2013).

The discussions of the PAM activities raised issues that had not been previously discussed by the members, such as good practice found in other MDTs. It became apparent that the workshop was the first time that the group had got together to discuss the functioning, as well as issues and problems, of the MDT meetings. This meant that the allocated time for discussions of their activities was not enough to complete the SSM process. The E's (performance criteria) were not discussed in detail and we did not arrive at a clear list of actions (the fourth methodological step for SSM) with an expanded description on how each should be done or agree to a timeline for each implementation. However, a number of actions to be taken were proposed when the PAM activities were explored.

All the outputs developed at the first workshop were distributed to all MDT stakeholders to ensure that the workshop discussion was recorded correctly and that the definitions and the model were agreeable to all the members. The aim was to enhance stakeholders' memory of any action resulting from discussing the models and hence make it possible for action to be realized. However, the responses to these outputs were limited due to the participants' time constraints and/or perhaps because they did not have any amendments or additions to make.

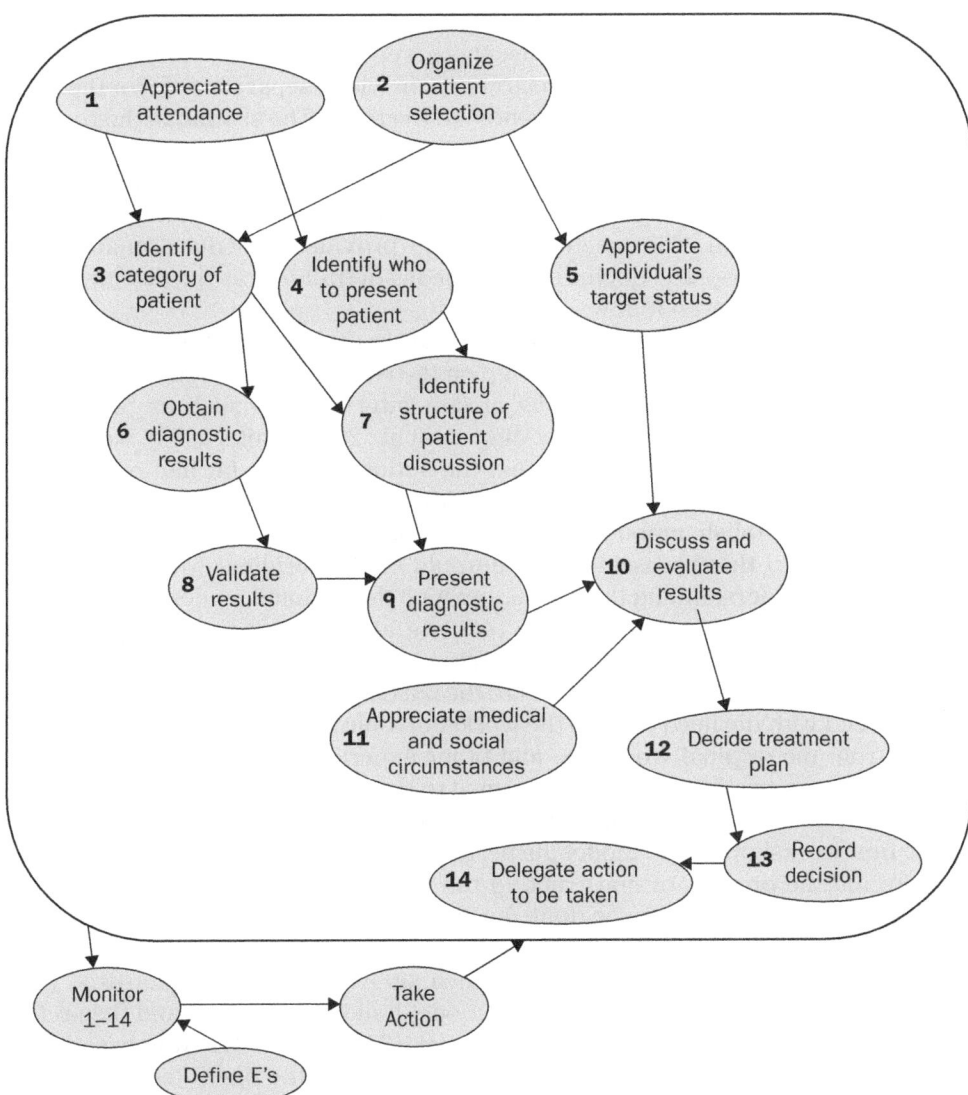

Figure 4.5 The MDT activities supporting the operational transformation
Source: Kotiadis et al. (2013).

Incorporating a performance evaluation element to the MDT

Observations of MDT (2009)

In 2009, the original analysts together with some new analysts approached the MDT to revisit the original intervention that ended shortly after the workshop took place in 2006. The aim was to establish if any action had been taken since the first workshop and to explore if developing a PMM in a workshop environment would be useful to the SSM process. A subset of the analysts observed the MDT in action and it was immediately clear that several improvements (actions), in line with the discussions of the 2006 workshop, had been implemented.

The MDT meeting had been officially timetabled, which meant that the team members were able to attend because there were fewer clashes with other duties. The meeting had moved to a much larger room with a separate area for the results of clinical investigations to be presented to the team. The seating in the room was arranged around a large U-shape table providing all members with the opportunity to listen and contribute to the discussions taking place. The team seemed to communicate effortlessly despite one of the key members communicating via video conferencing. The meeting now also provided an educational value to middle- and lower-grade doctors and surgeons in training, as the discussions and findings were easier to follow.

Having observed the radical improvement in the way the MDT was functioning, it was felt appropriate to follow up on the results of the original intervention. Unfortunately, because the results of soft interventions such as SSM are not easily quantifiable (Connell, 2001; Mingers et al., 2009; White, 2006), we could not establish or claim retrospectively that the action taken so far had resulted from the first workshop. Hence, we invited the MDT participants to a second workshop in order to establish, among other things, if any further action had taken place that related to the discussions and outputs of that original workshop in 2006. Apart from understanding the extent to which the notional model, represented by the PAM (Figure 4.5), had become real, we also wanted to judge it using the emerging idea of a PMM (Kotiadis, 2007).

Both workshops had been subject to the participants' strict time constraints and organized with the need to keep the workshops' duration to a minimum. On reflection from the original workshop and other experiences gained about clinicians' availability, a two-hour workshop seemed to be enticing to them. In this strict time slot, we had to allow time for both exploring the action that had taken place since the first workshop and to model the performance measurement aspects.

To aid the process of constructing the PMM and to speed up the process of explaining what needed to be done during the second workshop, we constructed a simple PMM (prior to the workshop) with few activities. Bringing preliminary models into a workshop is not a new idea, as Andersen and Richardson (1997) have also advocated their use in other model-building exercises and they refer to these as very stripped-down concept models. This basic PMM was based on the PAM (operational function) developed in 2006 during the first intervention.

Workshop 2 (2009): Debating and evaluating using the PMM

All the MDT members were invited to participate in the second workshop and around seven attended, which was most of the membership. The participants included surgeons, physicians, nurses, and coordinators. Workshop 2 was structured into six tasks. Task 1 was an individual brainstorming task (a pre-test) where each participant was asked to list three actions on a form that they thought should be undertaken to improve the performance of the MDT. The form was collected before proceeding to the next task. This task was also repeated towards the end of the workshop (referred to as Task 5), using a second form (a post-test), the aim of which was to assess the impact of the workshop process on their views (Kotiadis et al., 2013). Tasks 2 and 3 aimed at establishing whether the outputs constructed three years earlier (Table 4.1, Figure 4.5) were still relevant. More specifically,

Task 2 was to revisit the root definition and establish if this was still a reasonable definition of the MDT operations. Task 3 was to revisit the PAM diagram (Figure 4.5) for the operational aspect of the MDT and establish if there were activities that were not implemented and/or irrelevant, as three years had elapsed so we needed to establish the validity of the PAM.

The participants were very keen to discuss the MDT's activities. For example, the activity 'appreciate attendance' was of particular interest, as the histopathologist had some absences due to reductions in staffing (from 4.5 full-time equivalent down to 1) resulting in the cancelation of the MDT meeting on those occasions. The minimum number of attendants or specialties present to achieve quorum was discussed at length. Many ideas were put forward, including recording on their official documentation whether the MDT opinion on a case is 'valid or complete'. The educational benefit for trainee doctors and surgeons of specialist participation was also touted. A PAM diagram was found to be valid even though several years had elapsed.

Task 4 was to construct a PMM. A preliminary PMM with very few activities was shown to the participants in order to explain its structure. The workshop participants were told that the activities shown in the PMM were not necessarily correct but would give them an idea of how to construct it. Bringing a preliminary model into the workshop would also help in speeding up the process of constructing it. It should be noted that in our intervention, half the workshop time was dedicated to this task.

The participants were encouraged to consult the PAM and consider what information they would want to monitor for each activity or for a group of activities and what they would/could determine from that monitoring action or group of monitoring actions. The examples provided on the preliminary PMM acted as a starting point to the task but the participants soon engaged in conversation and were engrossed thereafter in relevant discussion without encouragement from the analysts. The analysts recorded the monitoring activities during the first part of this task with as few as possible interruptions so as not to hinder the trails of thought of the participants. Any interruptions to the conversation were to clarify the wording of the activities in the model. The 'monitoring' activities were then used to derive the 'determine if' activities. The PMM activities were then linked and reorganized after the workshop to ensure the minimum crossover of links. The PMM constructed as a result of the workshop can be seen in Figure 4.6.

Task 5 was to ask each of the participants to list three actions on a form (a post-test) that should be undertaken to improve the performance of the MDT. Task 5 also provided a warm-up exercise to Task 6 that was to get the participants collectively to discuss what action, if any, should be taken based on the PMM. However, many actions had already been partly discussed while constructing the PMM. The participants were subsequently asked to consider action that would be implemented in the short term (3–6 months) and longer term. The participants came up with three actions to be undertaken in the short term and were keen to create a system of evaluation that captured all the activities in the PMM for the longer term.

The MDT members took forward the action recorded as a result of the debate, and the analysts ended their involvement.

Figure 4.6 The performance measurement model supporting the operational transformation process of the MDT

Source: Kotiadis et al. (2013).

The main learning points to take from this case study are:

1. SSM offers a mode of engagement that can be applied to any problematic situation. In the case study, it enabled a multidisciplinary team to discuss their situation and come up with action to take forward as a group. SSM could be easily applied in developing countries to investigate the provision of basic health and social care, alleviating the effects of poverty (e.g. Hindle et al., 2015), etc.

2. SSM can be undertaken in workshops, with many stakeholders being given the chance to contribute and ultimately take responsibility for the actions to be taken. However, workshops require the prior organization and scheduling of the activities required for each stage of the methodology. In the case study described here, much effort was put into ensuring attendance at the workshops and the materials handed out were prepared in advance of the workshops.

3. SSM offers a low technological mode for a comprehensive investigation, leading to learning and determining further action. In our case study, we used word processing to print an agenda for our workshops and produce the SSM outputs from the tools used. However, the main ingredients to undertake the process were a few sheets of flip chart paper and marker pens. The main resource in SSM is the analyst, who must know the methodology and be able to use or guide others in using the tools.

Acknowledgements

I acknowledge the Daphne Jackson Trust and University of Kent for sponsorship. The case study presented has largely been taken from: Kotiadis, K., Tako, A.A., Rouwette and E.A.J.A. (2013) Using a model of the performance measures in Soft Systems Methodology (SSM) to take action: a case study in health care, *Journal of the Operational Research Society*, 64 (1): 125–37.

References

Andersen, D.F. and Richardson, P.G. (1997) Scripts for group model building, *System Dynamics Review*, 13 (2): 107–29.

Checkland, P. (1972) Towards a systems-based methodology for real-world problem solving, *Journal of Systems Engineering*, 3 (2): 87–116.

Checkland, P. (1999a) *Systems Thinking, Systems Practice*. Chichester: Wiley.

Checkland, P. (1999b) Soft systems methodology: a 30-year retrospective, in P. Checkland and J. Scholes, *Soft Systems Methodology in Action* (pp. A11–A15). Chichester: Wiley.

Connell, N.A.D. (2001) Evaluating soft OR: some reflections on an apparently 'unsuccessful' implementation using a Soft Systems Methodology (SSM) based approach, *Journal of the Operational Research Society*, 52 (2): 150–60.

Department of Health (1997) *The New NHS: Modern, Dependable*, Cm 3807. London: The Stationery Office.

Hindle, G., Vidgen, R., Hamflett, A. and Betts, G. (2015) *Business Modelling and Technology Leverage for Value Creation in the Food Bank Sector – Phase One Report*. Project report for EPSRC's NEMODE. [available at: http://www.nemode.ac.uk/wp-content/uploads/2016/03/Hindle-Foodbanks-phase-1-report-FINAL-18-12-15.pdf; accessed April 2016].

Kane, B. and Luz, S. (2011) Information sharing at multidisciplinary medical team meetings, *Group Decision and Negotiation*, 20 (4): 437–64.

Kotiadis, K. (2007) Using soft systems methodology to determine the simulation study objectives, *Journal of Simulation*, 1 (3): 215–22.

Kotiadis, K., Tako, A.A. and Rouwette, E.A.J.A. (2013) Using a model of the performance measures in Soft Systems Methodology (SSM) to take action: a case study in health care, *Journal of the Operational Research Society*, 64 (1): 125–37.

Mingers, J., Liu, W. and Meng, W. (2009) Using SSM to structure the identification of inputs and outputs in DEA, *Journal of the Operational Research Society*, 60 (2): 168–79.

Open Learn (n.d.) *Diagramming for Development 1 – Bounding realities*. Milton Keynes: The Open University [available at: http://www.open.edu/openlearn/science-maths-technology/computing-and-ict/systems-computer/diagramming-development-1-bounding-realities/content-section-3.1; accessed April 2016].

Pham, V. (2014) *Rich Pictures and CATWOE: Simple yet powerful scope-modelling techniques* [available at: https://elabor8.com.au/rich-pictures-and-catwoe-simple-yet-powerful-scope-modelling-techniques/; accessed April 2016].

Pidd, M. (2007) Making sure you tackle the right problem: linking hard and soft methods in simulation practice, in *Proceedings of the 39th Winter Simulation Conference: 40 years! The best is yet to come* (pp. 195–204). Piscataway, NJ: IEEE Press.

White, L. (2006) Evaluating problem-structuring methods: developing an approach to show the value and effectiveness of PSMs, *Journal of the Operational Research Society*, 57 (7): 842–55.

5 Cynefin: a tool for situating the problem in a sense-making framework

Annabelle Mark and Dave Snowden

Introduction: complexity and Cynefin

The field of complexity studies is comparatively new and there are few settled definitions and considerable differences between both theoretical and practical approaches, even within the humanities and social sciences (Byrne and Callaghan, 2013; Chandler, 2014). For many, complexity is seen as a subset of general systems thinking; for others, including the authors of this chapter, it represents a radical departure from thinking in terms of linear cause-and-effect relationships. The intention, therefore, is not to resolve those differences here, but to provide insight into the application of the Cynefin sense-making framework, which seeks to legitimize traditional cause-and-effect research models within boundaries. It is also to define new, distributed ethnographic approaches to handle truly complex systems where there can be no certainty and outcome-based approaches are *a priori* wrong. Thirdly, it is to explore the relationships between these differing perspectives. Sense-making as used here means: how do we make sense of the world so we can act in it? Inherent in that definition is a concept of sufficiency: how do we know enough to act?

Definition and history of the Cynefin framework

The Welsh word Cynefin (pronounced Kun-ev'in) is literally translated as habitat but has a wider sense of *the place of your multiple belongings*, a sense of being part of a flow of meaning over time that can never be fully understood, but that profoundly impacts on how we think and act. It was first published in its current form by Dave Snowden in 2002 and subsequently developed in collaboration with others (Kurtz and Snowden, 2003; Lazaroff and Snowden, 2006; Snowden and Boone, 2007) and more specifically in relation to health care (Mark and Snowden, 2006).

It seeks to combine action research with insights from complex adaptive systems theory, cognitive science, and anthropology. In operational use, the framework itself (see Figure 5.1) is an emergent property of people's understanding of exemplar narratives they express from their history (Kurtz and Snowden, 2003). The fluid shape of the framework emerges from the data, contrasting with a categorization approach where a model is pre-given and the data sorted into a predetermined structure. Such categorization can be efficient; but the problem is, if the context shifts between domains, or is at the boundary moving between

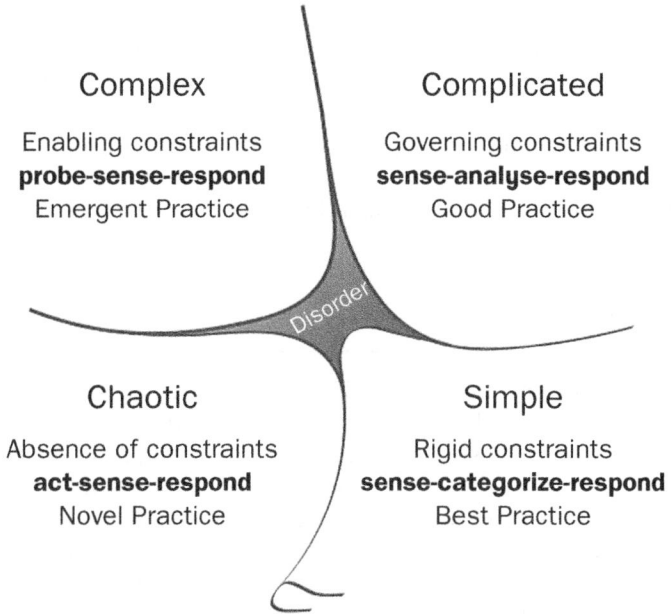

Figure 5.1 The Cynefin framework

domains, then it may result in categorization errors. Herein lies the second aspect of the framework, in that it is a dynamic model that acts as a metaphor for lived experience. The domains, and the boundaries between the domains, are defined by the narratives across a relevant research environment that may, for example, be an organization such as a hospital, or a whole community. These narratives are the primary sense-making mechanism by which we create common under-standings, the framework boundaries helping to distinguish between different types of action and analysis, necessary for human sense-making.

The boundaries also reflect distinct categories of systems:

- *ordered*, which can be simple or complicated, and
- *unordered*, which can be complex or chaotic.

These are further defined by the differing relationships between constraints and behaviour that may have to be resolved before the appropriate domain can be identified, and also take place within the Cynefin central domain of disorder (Fulop and Mark, 2013) when it is not clear which domain is appropriate. Constraints appear as both governing and enabling, defined as a rule and a heuristic (sometimes called a rule of thumb) respectively. For example, a rule will exist in each hospital about what colour scrubs are worn by different staff, denoting who they are and where they can go in theatre and beyond – these rules must be adhered to. However, a more general heuristic often unwritten or rule of thumb applicable anywhere for clinical staff across such institutions and environments in the UK is 'no jewellery below the elbow' on the wards or in theatre.

The purpose of the framework is to create a level of awareness in a decision-making community that involves first determining the nature of things in relation to the domains, then determining the nature of evidence possible and the type of interaction possible, within these contexts. As such, it can reduce conflict in both strategic and operational use by creating a language of legitimate differences. The domain determines the nature of what we do, and of how we understand the situation in front of us. If this seems a little abstract, the practical examples that follow will clarify the situation.

Cynefin: perspectives for health

The division between *ordered* as simple and complicated, and *unordered* as complex and chaotic, also helps us to understand the difference between mechanical and organic systems. And here is perhaps the best place to start, because health care is dominated by ordered systems in both its theoretical perspectives and its practice, yet it deals with the organic unordered, in the shape of the human body and the dominance of human behaviour, as its process. So the application of ordered systems, where the link between cause and effect is clear, is often not appropriate. For example, you would not apply the same rules to organizing a party (unordered) as you would to a project at work (ordered) (Snowden, 2005), becaues the former does not have outcomes we can manage or predict, only ways of managing the context (Snowden and Stanbridge, 2004), whereas most project management is predicated on an assumption of predictable outcomes. Exploring this helps in understanding the differences between ontologies, or ways of seeing our different worlds; it also helps in realizing that *unorder* is in fact a familiar space – which we do know how to manage, but differently.

Five domains

Ordered: simple

Here, cause-and-effect relationships are generally linear, empirical in nature, and not open to dispute; predictive models and the constraints of best practice are accepted such as epidemiological models and drug trials.

In research, the *decision model* is the domain of known cause and effect using, for example, quantitative techniques such as randomized control trials and statistical models.

Ordered: complicated

Cause-and-effect relationships in this domain are stable; though they may not be fully known, they can be discovered. The only issue is if we have access to the right expertise to investigate, sometimes over time and space. This also implies a key dependency on trust between the expert advisor and the decision-maker, and is the domain of systems thinking (as popularly understood as a discernible cause-and-effect relationship), but not complexity theory.

The *decision model* here is to sense incoming data, and analyse using relevant expertise. However 'entrained', sometimes expert or professional patterns are at their most problematic here, as they will only see what they go looking for through their own expert lens. A simple error in an assumption can lead to a false conclusion that is difficult to isolate. This kind of error was revealed most recently in 2009 during the UK Mid Staffordshire Hospital Enquiry (Francis, 2013), where an independent 2008 study into hospital standardized mortality ratios found that the 'Dr Foster' method used for calculations was prone to methodological bias (Mohammed et al., 2009). Other examples include the less tractable evidence of the social pressures on experts, especially doctors (Jureidini et al., 2003), or what Archie Cochrane, one of the fathers of evidence-based medicine, referred to as the 'God complex'. By this he meant, no matter how complex the problem, the person has an overwhelming belief that they are infallibly right in their solution.

These two ordered domains, of simple and complicated, are not based on individuals acting alone but on matters known to society, the organization or a profession; it is this collectivity that maintains the power and credibility of such shared perceptions.

Unordered: complex

This is the domain of complexity theory. Complex systems comprise many constantly interacting agents, such as the movement of a virus through the body, or the population. However, these patterns are not controlled by a directing intelligence; they are self-organizing systems making each illness or epidemic a unique and often unrepeatable pattern. The number of agents and the number of relationships defy categorization or analytic techniques. However, patterns that emerge through the interactions of many agents can be perceived after the event, as for example in the recent Ebola crisis (WHO Ebola Response Team, 2014), but cannot be predicted or made repeatable – within Cynefin this phenomenon is called *retrospective coherence.*

The *decision model* in this space is to create small parallel experiments that are *safe-to-fail* based on ideas of what might work to solve the problem in context. For example, in a project on aboriginal health and improved diets (Snowden, 2016b), over a dozen small experiments were constructed ranging from farming of desert products to ensuring the cultural provenance of killed meat. Over a period of time, those experiments merged and mutated to create a sustainable solution. The experiments came from the community who were engaged with the process. Conflict was radically reduced, as it was only necessary to agree that an experiment was coherent, not that it was right. Such experiments make the patterns or potential patterns more visible before taking any action. It is essential to sense the patterns and respond by stabilizing those patterns that are found to be desirable, and also destabilize those that are not desirable, and then *seed the space* so that patterns required are more likely to emerge. A good example of how *seeding the space* happens was shown within the pharmaceutical services in the UK, through the development of the role of primary care pharmacists. They carry out clinical and administrative work directly for family doctors and primary care organizations.

Following the economic liberalization of the NHS in the 1990s, a major stimulus for the growth of primary care pharmacy took place (Silcock et al., 2004) and effectively acted to *seed the space*, planting ideas about a different future. The establishment of the new professional group was not linked to a deliberate plan or change in health policy; but it was emergent, as a self-organizing system, because in complex systems such emergent phenomena can be catalysed but not designed.

Cynefin has developed a range of interventions designed to stimulate such emergence in complex knowledge interactions, for example (Snowden, 2004) to gain a better understanding of employee satisfaction with a more nuanced insight into cultural differences in the pharmaceutical industry.

What such methods also demonstrate is the important role of narrative as a valid research method for accessing this complex space, for example as a way of understanding the chaos presented to both patients and staff by critical illness and death (DelVecchio Good et al., 2004; Hold, 2015).

Unordered: chaos

In the chaotic domain there are no constraints or connections – the system is turbulent and there is no time available to investigate change as in many Fragile States (Mark and Jones, 2013). Fragile States here refer to countries that are failing or near to failure in many ways, including economically, although the interpretation of such definitions by donor states and organizations vary, adding further to the problem. This is because the primary concern of donors is to engage through a rational positivist approach (Mowles, 2011; Mowles et al., 2008) within geographic boundaries, rather than sense-making within the context of the whole environment or region, as analysed by Mark and Jones (2013), who explore these issues in the context of Cynefin.

If entered accidentally, the *decision model* in this space is to act, quickly and decisively, to reduce the turbulence as in the recent Ebola crisis (WHO Ebola Response Team, 2014), and then to sense immediately the reaction to that intervention, and respond accordingly. The trajectory of intervention will differ accordingly: it may be authoritarian intervention in response to a symmetric threat where the parameters are known and intention can be determined, or in the case of asymmetric threat, where the source and even the object of the intervention are unclear, there is a need to focus on multiple interventions, to create patterns moving towards the complex space. In health care, this can be the domain of the accident and emergency specialist responding to multiple failures in individuals, but also major accidents themselves, where the extent and effects are initially largely unknowable, as described by Shannon (2015). Key heuristics or rules of thumb, from professional knowledge rather than organizational sources, will guide much of the actions, enabling a shared professional response to context.

Table 5.1 summarizes the above four domains and responses.

Disorder – the fifth domain

The central area of disorder within Cynefin is less well understood (Kurtz and Snowden, 2003), but is the key to unlocking interpretation of the conflicts that

Table 5.1 Summary of four domains and responses

	NATURE	RESPONSE
CHAOS	**Unknowable unknowns** Temporary state – no time to analyse response No evidence of any constraints in operation High turbulence, no patterns ***Old certainties no longer apply***	**Act–Sense–Respond** Speed of authoritative response vital Follow and enforce heuristics prepared in advance Create constraints to allow solutions to emerge Use the opportunity to innovate
COMPLEX	**Unknown unknowns** Messily coherent, patterns discernible but not clear Constraints create patterns but not easily predictable Overall, the system exhibits flux within those constraints ***Evidence supports contradictory hypotheses for action***	**Probe–Sense–Respond** Create and monitor safe-to-fail experiments Ensure broad-range (often contradictory) experiments Increase/decrease constraints to test stability Agility of response key to amplify/dampen patterns
COMPLICATED	**Known unknowns** Ordered, predictable, forecastable Constraints evident and enforceable Stable and predictable within those constraints ***Evidence susceptible to analysis and expertise***	**Sense–Analyse–Respond** Determine experts or process to determine action Manage and enforce proper process Monitor effectiveness of constraints over time Focus on exploitation not exploration
SIMPLE	**Known knowns** Familiar, certain, well-worn pathways Constraints self-evident to all and accepted by all Very stable within universal constraints ***Self-evident solutions, little dissent or disobedience***	**Sense–Categorize–Respond** Ensure sound process in place and accepted Monitor for non-compliance Test for complacency in case constraints are excessive Protect some pet mavericks who challenge the system

exist among decision-makers in reaching agreement on the nature of a situation (Fulop and Mark, 2013). As a result, individuals compete on the basis of their preference for action. Those most comfortable with stable order will seek to create or enforce rules through control (simple). In health care, this can be experts, be they clinicians or managers, who may seek investment to conduct research to determine the 'right' answer (complicated).

In the unordered complex space, those acting politically, such as Trust Board members and sometimes managers, will increase the number and range of their contacts to increase communication and stimulate the emergence of a new solution (complex), as they intuitively understand the problems presented. Finally, in the chaotic domain, dictators benign or not, will use power, often without responsibility, to determine that a crisis has occurred, pushing the situation into chaos from which action and not thought is required, allowing their absolute control of the situation.

However, the reduction in size of the domain of disorder has ultimately to be a consensual act of collaboration; this is demonstrated in any emergency where clinical teams are required to work together. The *decision process* in this, the most unusual of the five domains, must be relational, involving the co-construction of context through the processes of organizing (Fulop and Mark, 2013). The positive aspect of the domain is that it remains a legitimate transition space for innovation.

Four dynamics

The framework is not confined to classification by domains but also the movement between them. The reality of complex systems in particular is that events move between domains, and actions can retard, enable, and accelerate those changes. Although many dynamics are possible (Kurtz and Snowden, 2003) there are four dominant ones.

The principal and most resilient dynamic within Cynefin is known as the *operational dynamic* (shown as within the solid line circle in Figure 5.2) and represents a natural cadence between complex and complicated. As patterns emerge in the complex domain, through parallel *safe-to-fail* experiments used to explore a situation knowing we can fail without damage, we attempt to increase the level of constraint through rules or rules of thumb. If that produces increasing predictability of outcome, then we can shift from complex (exploration) to complicated (exploitation). What is often missing from this common practice is constant testing to determine whether the constraints are still producing predictable results. If they are not, then we need to shift back into the parallel *safe-to-fail* experimentation of the complex domain. The failure to do this can result in 'best' becoming 'past' practice, for example requiring day care eye surgery patients to remove all their clothes, and as such can also result in catastrophic failure (if patients fail to attend the eye clinic). Thus, too rigid controls on safety can often result in *work arounds* (or avoidance of controls) being necessary for the job to be completed. Rule breaking like this then becomes habitual and accidents are more likely to happen. The natural cadence of this dynamic will change according to context; in some cases the cycle time will be years, in others months or less.

Shifting from complicated to simple (the *standardization dynamic* shown as the arrow in Figure 5.2) is done when there is little likelihood of a need to change in the medium- to long-term prognosis. Once shifted to the simple domain, there is no return – only abandonment, sometimes into chaos, so it should not be done lightly. One of the dangers of process-led change initiatives, such as

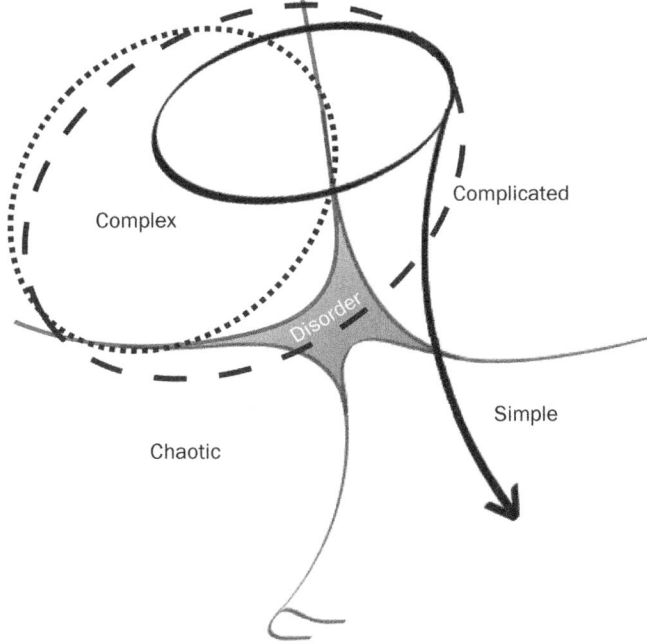

Figure 5.2 Cynefin with dynamics

Lean Management (Radnor et al., 2012), is an over-focus on this domain at the cost of organizational resilience and leadership credibility.

The *innovation dynamic* (shown as within the dashed line circle in Figure 5.2) works over longer cycles than the *operational dynamic*. As a body of knowledge becomes more and more the domain of experts, there is increasing likelihood of *entrained thinking* over the longer term leading to a narrowing perspective. This requires a cyclical process of disruption in which expert knowledge is taken on a shallow dive into chaos through a near complete relaxation of constraint, to allow a novel pattern or patterns to evolve which shifts into the complex domain. One example of this is the use of a completely different discipline to look at common practice within a professional group. In practice for Cynefin, this has involved, for example, using a liturgical design specialist to rework information transfer in operating theatres in a hospital in Bellingham, Washington state, because liturgical design is about compressing complex ideas into simple memorable form, challenging the existing over bureaucratic norms.

Finally, we have the *grazing dynamic* (shown as within the dotted line circle in Figure 5.2), in which the overall volatility is such that stability is rarely other than transitory and we constantly skim the surface of chaos. This is increasingly common in public health, consumer goods, and social computing. It requires a very different approach to management, one that goes beyond *safe-to-fail experiments*, which we use to explore a situation knowing we can fail without damage. Interventions instead must become smaller, faster, and more distributed with real time feedback loops.

Method

Overview

The Cynefin framework brings together a set of ideas making the interaction of differing perspectives coherent but not complementary (Snowden, 2002). It provides new ways of seeing and integrating ideas, but its development, to understand the complex space in particular within the framework, is found within the research instrument SenseMaker®, providing the capacity to undertake large volume studies in diverse contexts, using either hand-held devices and apps or websites (www.cognitive-edge.com, SenseMaker tab), for example in capturing the experience of a patient's journey through the health system. This is done by significant numbers of participants keeping a diary or record of observations of their condition and the surrounding environment, but uniquely includes self-interpretation of the data. In the complex domain of Cynefin, there are few if any linear relationships between cause and effect; however, we can make statements about the propensities of aspects of the system, and how it might evolve and change. So while we can measure the current state of the system, we cannot make accurate statements about the future. However, the ability of humans to impute cause and effect with the benefit of hindsight using *retrospective coherence* increases our problem in creating evidence that can lead to action. Current dispositional states that may infer possible future actions and possible future states can now also be mapped based on the development of 'fitness landscapes' (Kauffman, 1997), which are applicable to mass data capture needs such as public health.

Source or primary data

Capturing data in various forms and interpreting it is at the heart of the process, but what changes in Cynefin is who decides on the questions, and how interpretation is most likely a collaborative activity.

The goal of research and monitoring here is to gather the micro-narratives or observations of day-to-day experience in the complex space. To understand culture, it is better to listen in on informal conversations than to gather evaluative stories in a workshop. Material from the latter is profoundly influenced by the fact that the participants are aware of the purpose of the exercise and will respond accordingly. At workshops, interviews can only take place periodically, so lack the real-time feedback loops. In a patient journey, it is the mundane hour-by-hour experiences that need to be captured, not what is reported days or weeks later within a facilitated environment.

So the approach seeks to gather these micro-narratives or observations as close to the experience as possible, by enabling the recall of an experience without revealing a hypothesis or purpose in a survey environment. For example, in the annual event of the staff feedback survey, we ask questions such as '*Does your manager consult you on a regular basis?*' with a request for a response on a scale from 0 ('not at all') to 10 ('all the time'). The question contains an implied hypothesis that consultative practice by managers is good, and the manager's behaviour is being evaluated on a scale between good and bad. Where people know what answer you want, it influences their response.

In the Cynefin approach, we try to adopt a non-hypothesis-promoting question often in collaboration with those being researched, thus for instance: '*What story would you tell your best friend if they were offered a job in your workgroup?*' The response to that question, which contains no hypothesis, is then written by the user, or recorded or represented by a picture (or any combination of the three).

This can be done via a smart phone or tablet as well as on a website. The resulting combination is then signified into a quantitative framework (something we will describe in the next section) by the person who provided the material. Critically, they are interpreting the narrative, rather than evaluating their manager's behaviour. The material can be collected through a variety of media sources and is held in its original form because, for example, tone of voice can be very important.

When respondents are keeping a journal, a prompting question, such as the one indicated above, is not needed. For example, on current work in NHS Wales on patient journeys, the respondent is asked to make an observation at the start and end of each day, at specific events, or when they feel some change in their condition. The exact instructions vary a little by project but the basic principle is a non-hypothesis question in which the respondent cannot decide what the right answer is to please the organization they work for. We also increase the volume of material and it can be received in real time by decision-makers at all levels. This is why kiosk approaches work well in situations such as A&E admission or primary care waiting rooms. Alternatively, most notably within a mental health project in Northern Ireland, the carer or a relative (who may also be the carer) in effect interviews the subject, and assists them in interpretation. Bar codes can also be placed at physical locations so that the patient scans the code before contributing their observation or micro-narrative. These also act as prompts in their own right and additional instructions or requests can be co-located with the codes.

Signification

Key to the Cynefin approach is the self-interpretation (signification) of the material at the time and point of collection by the research subject. The power of interpretation is transferred to the subject and away from the researcher. But we do not want a single *wrong* or *right answer* to be visible – we want to attempt to move towards an understanding of a system as non-causal or dispositional, that is with no linear cause-and-effect relationships. In addition, we also need real-time feedback loops if we are to be authentic to the nature of complex adaptive systems.

After having told a story, taken a picture, or whatever the individual provided, the material is presented with a series of *triads* (normally six maximum) into which they interpret their contribution. Let us illustrate this as follows.

One of the triads will have the question '*Was the manager's behaviour . . .*' and then the triad aspects may be *altruistic, assertive, analytical*. The respondent balances the three aspects against each other, putting a point to show this within the triad (as shown in Figure 5.3). All labels are positive, so no indication is given of what is desirable or undesirable. Aside from preventing the inevitable bias that

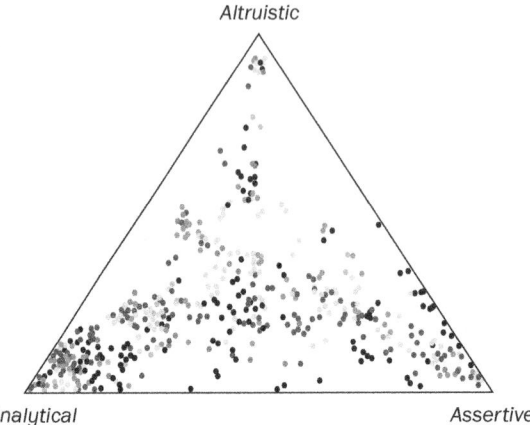

Figure 5.3 Colour coding by demographics. Each dot represents a story or picture that can be interrogated to explain the numbers

comes with knowing what answer is wanted, this approach also reflects the reality of having to balance different qualities. Given a choice, all three should be at 100% and with three linear scales that could be the response. With the triads they have to weight preferences.

Triads in SenseMaker® are descriptive and not evaluative. A second type of signifier is more evaluative. For example, on patient journey projects, where it is important to understand pain on a minute-by-minute basis, the respondent (or their carer) is presented with a grid that contrasts empathy with pain on a low to high scale in each case. The respondent then places three *stones* onto the grid and the stones are labelled '*As I see it*', '*As I think the nurse sees it*', and '*As I think the doctor sees it*' respectively. Those three stones show different perspectives.

Ideally, the signifier labels are derived from the field in a number of ways. For instance, in a patient safety project, the literature search identified nine key concepts or potential drivers. Where those drivers represented a trade-off, they were combined on a single triad, while the concepts had a triad to themselves, with three aspects indicated. Participants are actively engaged in designing their own signifiers, and the results are always carefully tested with representative groups before going live. Triads developed this way can be used again with the same groups in longitudinal studies, or across other organizations as in a comparative study.

So we have material gathered in various media, which is then interpreted at the point of origin into a quantitative framework, both descriptive and evaluative in nature. But the interpretation is of the observation – it is concrete and situated in that underlying narrative rather than being an abstract interpretation of partly remembered experience. From this, we can look at underlying patterns; the data can also be extracted to standard statistical and visualization tools within Sense-Maker®. Critically, once a pattern is seen the underlying narrative can be recalled by simply clicking through on electronically generated media.

Strengths and weaknesses

One of the most common comments when people are engaged in their first project is to suggest that the respondents are not interpreting their narratives correctly. In reality, no one can interpret their own narrative incorrectly; if you don't understand the interpretations, then you are seeing their world differently. Stories they saw as negative turn out to be positive from a respondent's point of view and vice versa. One basic technique to prepare people for this is to ask all members of the team to interpret a common story the way they think it would have been interpreted, by the person who told it. The differences are then discussed and at that point they are better prepared for differences, which then become a major strength of the approach. The fact that the technique is still novel means that preparation is key.

One way to explain it is to compare it with traditional longitudinal research. Here, narrative material is captured and transcribed. It is then interpreted by text analysis, computer algorithm or the researcher into a 'theory of the field'. In the case of SenseMaker®, the theory of the field is converted to signifiers. SenseMaker® then has the following advantages over traditional techniques:

1. Because the material is self-interpreted, there is no need for transcription and visual images can also be captured, overcoming language or disability barriers.
2. For the same reason, originating material can be in any language, only the signifiers have to be translated.
3. Self-signification means there is no time interval between capture and analysis, allowing for monitoring as well as research and weak signal detection of emerging trends.

But it can take time for people to realize the difference. So experimentation, use by the research team, early partial analysis, and testing are all desirable – as is a marketing plan to gain participation. There are no magic bullets in that respect. Taking the patient journey example mentioned earlier, real-time feedback allows early detection of safety and attitudinal issues and lower-risk early interventions In a series of projects in Vietnam and Columbia relating to subsistence crops and small businesses in slum areas respectively, real-time monitoring has allowed for prompt intervention at an early stage, reducing cost and also preventing negative tropes from emerging (Snowden, 2016a).

In comparative work in Northern Ireland, Wales, and two NHS Boards in England, the approach taken was to design six experiments in different fields of health care in parallel. This was done to reduce cost, but also to allow experiments that could fail, as some in the portfolio could then proceed. One major criticism of pilots is that too often they are not allowed to fail, and hence they provide less opportunity for learning. The importance of ensuring that you can learn from an intervention has been emphasized in Snowden's concept of *safe-fail experiments* (Snowden, 2010): the small interventions in which you first elicit ideas for tackling the problem from anyone who has one, and design safe-fail experiments to test each. Next, flesh them out, cost them, and subject them to challenge and review – with the aim of keeping experiments small but carrying out a broad range of them. Crucially, for each experiment to be valid, it should be possible to observe whether results are consistent with the idea or whether the idea is proved wrong. Another suggestion is to seek surprise and look into events and outcomes that were wholly unexpected at the outset of an intervention (Guijt, 2008).

One major advantage of the method is that it combines the objectivity and democracy of numbers with the explanatory and pervasive character of anecdotes. The method thus builds advocacy into the system, increasing the chances of change. In addition, the raw material is never abandoned. Once the report is written, it remains available to see patterns over longer periods.

Data analysis and management

Type of data generated

For triads, the system holds the position of the interpretation within the triad as well as the area allocated to a label expressed as a proportion of the whole, so each triad becomes three scales. The signification is always linked to the originating observation or story. This allows for three types of interpretation and analysis:

1. The most basic, as well as the most useful in many cases, is simply to present the data back as distribution or heat maps with some description (see Figure 5.4). This is normally done on an android device and the users can recall a triad, see the pattern, and select an area which then provides a list of narratives/observations to read, listen to, or view by clicking through the relevant dots.
2. As stated, the data is tested for correlations and cross tabs, either within the software or via export to most statistical packages. At this point from a statistical point of view, there are no major differences with other techniques other than the ability to move directly to the source material.
3. Finally, developing the use of fitness landscapes (Kauffman, 1997) can provide the means to integrate huge volumes of data into a simple visual form. We can see the pattern for the entire hospital or for a ward all from the same source data. This 'fractal' representation of a dynamic system also allows for contextually appropriate intervention design using the 'more stories like these, fewer like those' approach to managing the evidence generated.

Ideally, the material is made available to the participants as well as managers and researchers for their own interpretation and determination of actions.

Cleaning data

Permissions can be granted at the level of an observation/narrative. Typically, the person creating the material is given options: (i) anyone can see this, (ii) anyone with the same condition, (iii) only a researcher, or (iv) delete it. The permission here relates to the observation or narrative being interpreted. The metadata, or signification, contains no confidential data and is always captured and retained. If most of the material that would explain a statistical pattern has been deleted, that is a message in its own right!

In addition, removal of personal or place names can be automated, as can key words to trigger medical review so maintaining the ethical requirements within health research.

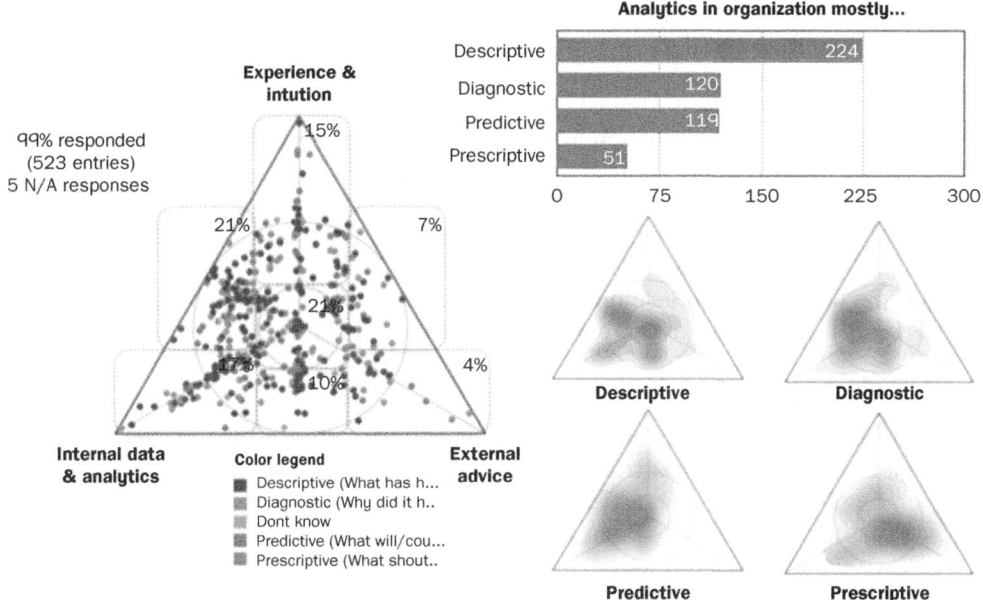

Figure 5.4 Most decisions rely on a mixture of Experience & intuition and Internal data & analytics, but organizations that make use of Prescriptive analysis rely more heavily on External advice

A new theory of change, from outcomes to vectors

The real change in organizations or communities is when you change the way that people connect, best achieved through small actions that change perceptions in an evolutionary way. If you want systemic change, it is a lot easier to change the interactions through individualized as well as mass evidence, as generated through Cynefin, allowing people autonomy over what they are and what they might become.

With SenseMaker®, the ability to ask in real time '*What would we do tomorrow to create more like these, and fewer like those*', allows wider engagement over different educational, literacy, and experience levels. It also allows the setting of vector or directional targets to shift the pattern very broadly to an *adjacent possible*. To achieve that change, the experience of people recording data and the way they interpret it has to change as this approach does; in addition, it is also far more difficult to 'game' or fix than outcome or output targets.

Cases studies

10,000 Voices

Why

The 10,000 Voices Initiative is commissioned and funded by the Public Health Agency (PHA) and the Health and Social Care Board in Northern Ireland. The name of the initiative is a metaphor for listening to patients', their carers' and families' voices on a large scale, collecting their experiences in the form of micro-narratives mapped on to purpose-built signification frameworks. The intention of this initiative is not only to 'see the person behind the data' but also to understand how the aggregated data can affect and influence the way services are commissioned and delivered (Armstrong, 2016). As a result of the information received, a number of improvements have been identified and are being progressed to enhance the experience for patients/clients and family members and to influence how services are shaped in the future.

Through 10,000 Voices, a regional structure has been developed to capture, understand, and improve patient and client experience using a blended approach of qualitative and quantitative information. This is based on a partnership approach, which is underpinned by the principles of Experience Based Co-design (NHS Institute for Innovation and Improvement, 2011).

How did they do it?

This initiative has been implemented in collaboration with the six Health and Social Care Trusts in Northern Ireland, with a focus on specific areas of care, including unscheduled care, care at home, and nursing and midwifery care.

Trust facilitators have been appointed to develop standard processes for data collection, analysis, reporting, and where necessary facilitating quality improvement plans. Comprehensive regional and local engagement and communication plans are developed by the Trust facilitators and are adapted as necessary for specific areas of health and social care. Standard processes include the provision of SenseMaker® software, electronic (mobile app and computer) and paper formats, with the ability to type or record micro-narratives as desired. The story template was translated into six languages: Traditional Chinese, Simplified Chinese, Latvian, Slovakian, Lithuanian, and Polish.

The analysis stage has a two-pronged approach. Micro-narratives are reviewed on a weekly basis by the Health and Social Care Trusts and Public Health Agency. This allows for patterns in the data to be identified and action taken in a timely way by the teams within an agreed protocol. In addition, service users and staff are brought together in facilitated workshops, which provides opportunities to analyse and interpret the data together, resulting in further local ownership in terms of contextualized local action plans. Areas requiring regional action are identified and shared with the appropriate staff so that actions can be progressed as necessary.

Figures 5.5–5.8 depict four triads representing some of the questions and responses given by different groups of respondents.

What issues were encountered?

One Trust was unable to recruit a trust project facilitator, which resulted in administrative staff being recruited solely to collect data so as not to cause any delay in the collection phase (see Figure 5.8).

What did they get out of it?

10,000 Voices has helped to raise the profile of patient and client experience in Northern Ireland, and provides a rich source of information from over 6000 patients that has been used to influence the delivery of services, inform practice, regional policies, and training and education (Donaldson et al., 2014).

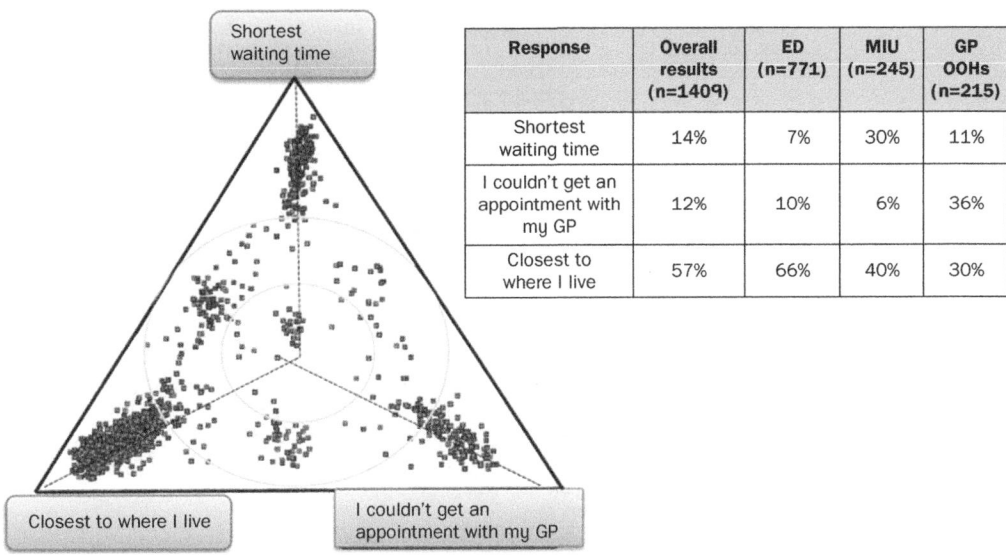

Response	Overall results (n=1409)	ED (n=771)	MIU (n=245)	GP OOHs (n=215)
Shortest waiting time	14%	7%	30%	11%
I couldn't get an appointment with my GP	12%	10%	6%	36%
Closest to where I live	57%	66%	40%	30%

Figure 5.5 Question 1: 'What made you decide where to go for help?'

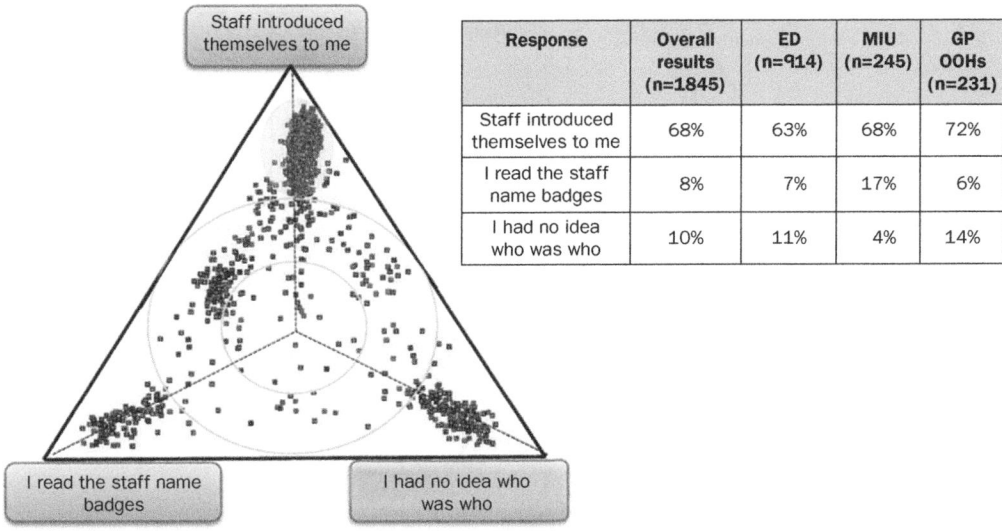

Response	Overall results (n=1845)	ED (n=914)	MIU (n=245)	GP OOHs (n=231)
Staff introduced themselves to me	68%	63%	68%	72%
I read the staff name badges	8%	7%	17%	6%
I had no idea who was who	10%	11%	4%	14%

Figure 5.6 Question 2: 'How did you know who was looking after you?'

Flourishing Communities Index

Why

The Catholic Organization for Relief and Development Aid (Rutten, 2015), head-quartered in The Hague with country offices in 11 countries, 'fights poverty and exclusion in the world's most fragile societies and conflict-stricken areas'. This award-winning assessment initiative, the Flourishing Communities Index (FCI), developed around the idea of giving local communities a global voice through

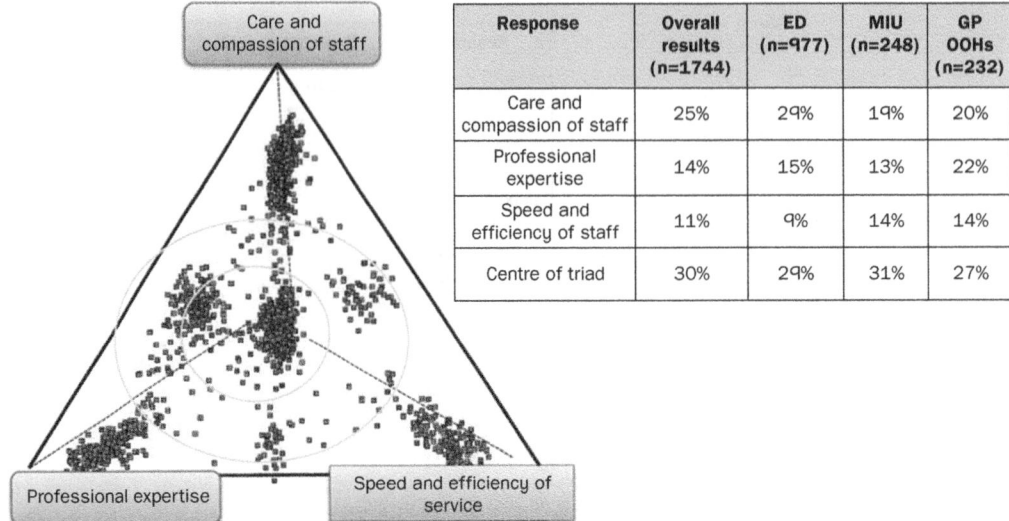

Response	Overall results (n=1744)	ED (n=977)	MIU (n=248)	GP OOHs (n=232)
Care and compassion of staff	25%	29%	19%	20%
Professional expertise	14%	15%	13%	22%
Speed and efficiency of staff	11%	9%	14%	14%
Centre of triad	30%	29%	31%	27%

Figure 5.7 Question 11: 'Overall, what were you most satisfied with?'

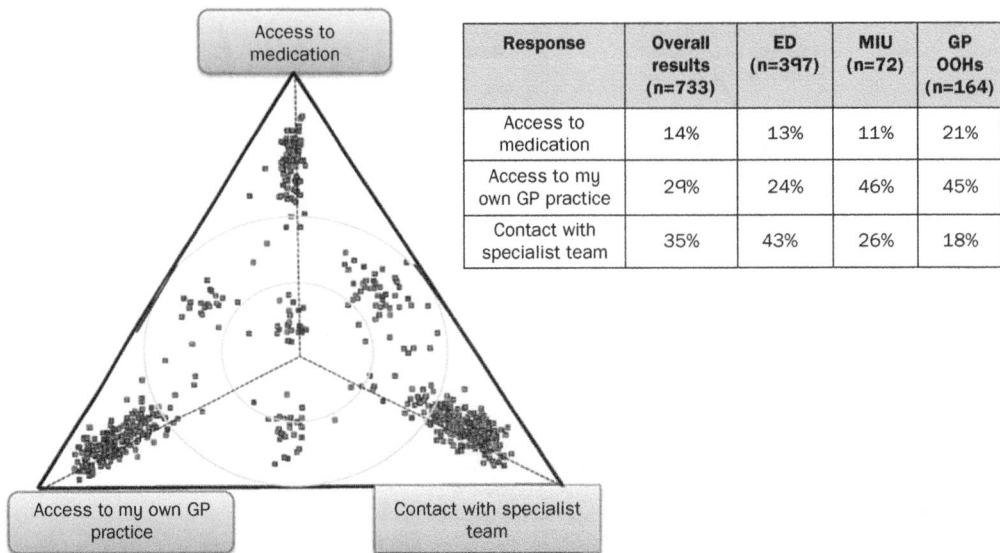

Response	Overall results (n=733)	ED (n=397)	MIU (n=72)	GP OOHs (n=164)
Access to medication	14%	13%	11%	21%
Access to my own GP practice	29%	24%	46%	45%
Contact with specialist team	35%	43%	26%	18%

Figure 5.8 Question 13: 'What would have enabled/supported you to stay at home?'

a rigorous framework that 'meets the demand from academics, development communities and donors for impact evaluation methods that are sensitive to local contexts and social difference' (Rutten, 2015).

How did they do it?

The initiative gathered information from a representative selection of community members (at least 200) by using trained story collectors, each equipped with the

SenseMaker® app. After sharing their story, mostly through the audio recording function of the app, the respondents were then asked to interpret, or signify, their story on a series of triads developed together with Cordaid. The testing phase saw the design tested in six countries, with almost 100 stories collected. The SenseMaker® signification framework was then further refined based on the results of the test phase and general feedback, including a sense-making workshop to analyse the data. The FCI is now ready for a pilot in the Democratic Republic of Congo.

What issues were encountered?

Issues were broadly divided into two categories – methodological and technical, with the latter involving uploading, transcription, and translation of audio narratives. Methodological issues arose in the ethnographic process, including the preparation and training of story collectors and how best to guide sensitive topics. In some cases, some signifiers were perceived as being too abstract.

What did they get out of it?

As a start, data from a Burundi community provides insights for Cordaid's programming and lobby and advocacy work at local and national levels. A theme to emerge was health, which showed a positive perspective on family planning and influenced Cordaid's healthcare programming with a focus on sexual and reproductive health, and lobbying for family planning (Figure 5.9).

Another theme that emerged, under 'Other', was about land conflicts due to a corrupt legal system. Cordaid was then able to start programming for security and justice, and addressing corruption in the legal system (Figure 5.10).

The triad data in Figure 5.10 indicate that many of the stories were driven either from communities or authorities, with hardly a private sector of any importance. This finding is coherent with the data in Figure 5.11.

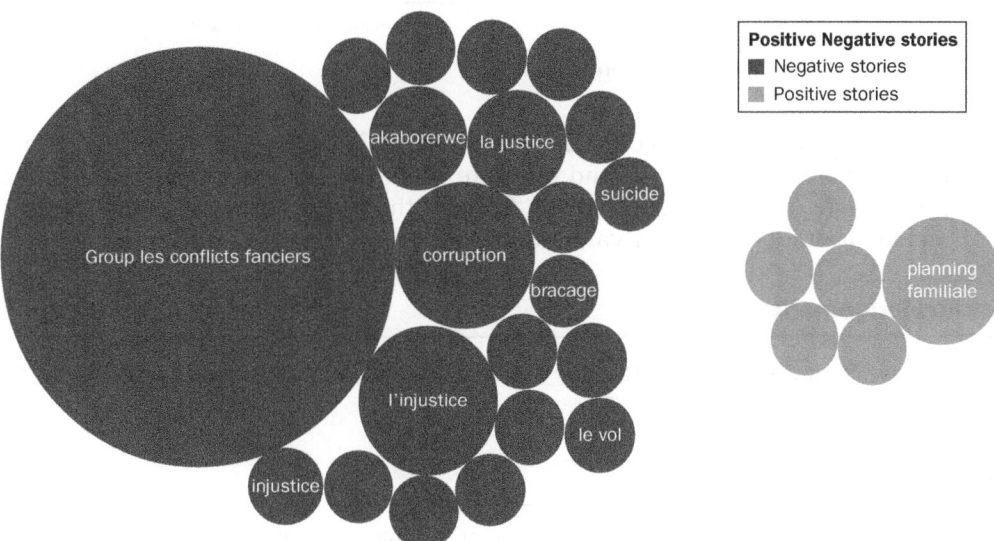

Figure 5.9 Other themes grouped versus Emotional tone

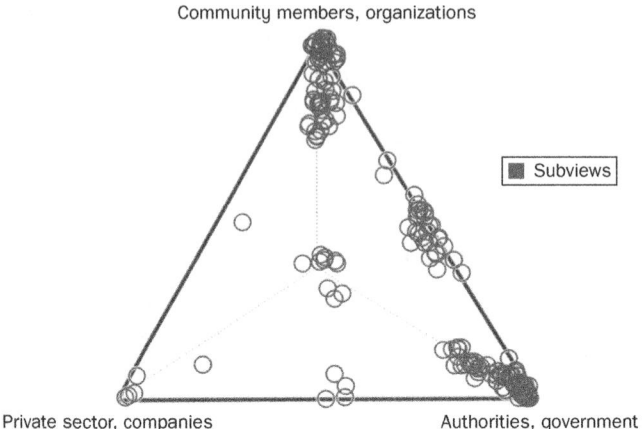

Figure 5.10 'Who took action in your story?'

Figure 5.11 'Who should take action in your story?'

The test phase allowed Cordaid to adjust the FCI framework accordingly, fine-tune formats and methods, and estimate costs for implementing it as a global programme. This initiative has won Cordaid the best-of-show award at the 2015 annual meeting of the UK Evaluation Society.

Acknowledgement

The work of Jules Yim of Bangor University in assembling the case material for this chapter is acknowledged.

References

Armstrong, C. (2016) *Seeing the Person Behind the Data*. Belfast: Public Health Agency [available at: http://www.kingsfund.org.uk/sites/files/kf/media/Christine_Armstrong.pdf; accessed 12 May 2016].

Byrne, D. and Callaghan, G. (2013) *Complexity Theory and the Social Sciences*. Abingdon: Routledge.

Chandler, D. (2014) *Resilience: The governance of complexity*. Abingdon: Routledge.

DelVecchio Good, M.J., Gadmer, N.M., Ruopp, P., Lakoma, M., Sullivan, A.M., Redinbaugh, E. et al. (2004) Narrative nuances on good and bad deaths: internists' tales from high-technology work places, *Social Science and Medicine*, 58 (5): 939–53.

Donaldson, L., Rutter, P. and Henderson, M. (2014) *The Right Time, the Right Place: An expert examination of the application of health and social care governance arrangements for ensuring the quality of care provision in Northern Ireland* [available at: http://www.belfasttrust. hscni.net/pdf/The_right_time_the_right_place_DOC.pdf; accessed 14 July 2017].

Francis, R. (2013) *Report of the Mid Staffordshire NHS Foundation Trust Public Inquiry*. London: The Stationery Office [available at: https://www.gov.uk/government/publications/ report-of-the-mid-staffordshire-nhs-foundation-trust-public-inquiry; accessed 14 July 2017].

Fulop, L. and Mark, A. (2013) Relational leadership, decision-making and the messiness of context in healthcare, *Leadership*, 9 (2): 254–77.

Guijt, I. (2008) *Seeking Surprise: Rethinking monitoring for collective learning in rural resource management*, PhD thesis, Wageningen University, Wageningen, The Netherlands [available at: http://edepot.wur.nl/139860].

Hold, J.L. (2015) A good death: narratives of experiential nursing ethics, *Nursing Ethics*, 24 (1): 9–19.

Jureidini, J.N., Shafer, A.T. and Donald, T.G. (2003) 'Munchausen by proxy syndrome': not only pathological parenting but also problematic doctoring?, *Medical Journal of Australia*, 178 (3): 130–2.

Kauffman, S. (1997) *At Home in the Universe*. Oxford: Oxford University Press.

Kurtz, C. and Snowden, D.J. (2003) The new dynamics of strategy: sense-making in a complex and complicated world, *IBM Systems Journal*, 42 (3): 462–84.

Lazaroff, M. and Snowden, D. (2006) Anticipatory modes for counter-terrorism, in R. Popp and J. Yen (eds.) *Emergent Information Technologies and Enabling Policies for Counter-Terrorism* (pp. 51–73). Hoboken, NJ: Wiley-IEEE Press.

Mark, A. and Jones, M. (2013) Thinking through health capacity development in Fragile States. *International Journal of Health Planning and Management*, 28, (3) 269–289.

Mark, A. and Snowden, D. (2006) Researching practice or practising research – the contribution of Cynefin, in A. Casebeer, A. Harrison and A. Mark (eds.) *Innovation in Healthcare: A reality check* (pp. 30–41). Basingstoke: Palgrave Macmillan.

Mohammed, M.A., Deeks, J.J., Girling, A., Rudge, G., Carmalt, M., Stevens, A.J. et al. (2009) Evidence of methodological bias in hospital standardised mortality ratios: retrospective database study of English hospitals, *British Medical Journal*, 338: b780 [doi: 10.1136/bmj.b780].

Mowles, C. (2011) Post-foundational development management – power, politics and complexity, *Public Administration and Development*, 30 (2): 149–58.

Mowles, C., Stacey, R. and Griffin, D. (2008) What contribution can insights from the complexity sciences make to the theory and practice of development management?, *Journal of International Development*, 20 (6): 804–20.

NHS Institute for Innovation and Improvement (2011) *Method, Toolkit and DVD 'how to' use EBCD*. Coventry: NHS Institute for Innovation and Improvement.

Radnor, Z.J., Holweg, M. and Waring, J. (2012) Lean in healthcare: the unfilled promise?, *Social Science and Medicine*, 73 (3): 364–71.

Rutten, R. (2015) Flourishing Communities Index: Care. Act. Share. Like Cordaid. Cordaid Factsheet, May [available at: https://www.cordaid.org/en/wp-content/uploads/sites/3/2015/06/ Cordaid-10135-04-FCI-4pager-DEF-LR.pdf; accessed 14 July 2017].

Shannon, T. (2015) Cynefin and healthcare, *Fractally Speaking*, Cynefin UK.

Silcock, J., Raynor, T.D.K. and Petty, D. (2004) The organisation and development of primary care pharmacy in the United Kingdom, *Health Policy*, 67 (2): 207–15.

Snowden, D. (2002) Complex acts of knowing: paradox and descriptive self-awareness, *Journal of Knowledge Management*, 6 (2): 100–10.

Snowden, D. (2004) Narrative patterns: the perils and possibilities of using story in organisations, in E. Lesser and L. Prusak (eds.) *Creating Value with Knowledge* (pp. 201–16). Oxford: Oxford University Press.

Snowden, D.J. (2005) Multi-ontology sensemaking – a new simplicity in decision making, *Informatics in Primary Care*, 13 (1): 45–54.

Snowden, D. (2010) Naturalizing sensemaking, in K.L. Mosier and U.M. Fischer (eds.) *Informed by Knowledge: Expert performance in complex situations* (pp. 223–34). New York: Psychology Press.

Snowden, D. (2016a) *On Stony Ground* [available at: http://cognitive-edge.com/blog/on-stony-ground/; Blog]. Cognitive Edge.

Snowden, D. (2016b) *The Hound Calls* [available at: http://cognitive-edge.com/blog/the-hound-calls/; Blog]. Cognitive Edge.

Snowden, D. and Boone, M. (2007) A leader's framework for decision making, *Harvard Business Review*, 85 (11): 68–76.

Snowden, D. and Stanbridge, P. (2004) The landscape of management: creating the context for understanding social complexity, *E:CO Special Double Issue*, 6 (1/2): 140–8.

WHO Ebola Response Team (2014) West African Ebola epidemic after one year slowing but not yet under control, *New England Journal of Medicine*, 372 (6): 584–7.

6 | Causal loop diagrams: a tool for visualizing emergent system behaviour

Andrada Tomoaia-Cotisel, Hyunjung Kim, Samuel D. Allen and Karl Blanchet

The mere formulation of a problem is almost half of its solution.

David Hume (1711–1776)[1]

Policy-makers are better able to identify and implement effective health system strengthening (HSS) efforts when they have an adequate understanding of the structure and emergent behaviour of the health system they are attempting to strengthen. Without this, decisions can result in unintended consequences or encounter policy resistance where ill effects outweigh the intended benefits (de Savigny and Adam, 2009; Sterman, 2006). Nevertheless, a comprehensive review of 106 HSS evaluations found that although many referred to a conceptual framework and considered impacts beyond the targeted part of the health system, none considered non-linearity of effects, delays or interactions between different system parts (Adam and de Savigny, 2012). The field of system dynamics modelling[2] has developed a rigorous methodology for eliciting, documenting, and visualizing these factors: causal loop diagrams (CLDs).

This chapter presents a definition of CLDs, a description of ways in which they are used, standards for their visualization, and case examples of the methodologies for eliciting and using them. We conclude with a discussion of where CLDs fit within a broader study of a complex system and refer to resources for further training.

Causal loop diagrams

A CLD is a visual representation of a dynamic hypothesis, and consists of causal linkages among elements of a system (or 'system structures') thought, over time, to generate a specific problem ('emergent behaviour').[3]

CLDs are one of a range of visual representation tools, generally referred to as 'mental maps' (Huff, 1990). An earlier approach to visualizing mental maps can be found in Axelrod (1976), where he analysed statements made by political elites and represented their causal assertions in words-and-arrow relationships. Similar methods include Eden and Ackermann's cognitive maps (Eden et al., 1992) and Bougon's congregate cognitive maps (Bougon, 1992).

Uses of CLDs

CLDs are used in health systems research and policy-making to:

Communicate mental models. Decision-makers have 'mental models' of the system that they attempt to manage. These mental models represent decision-makers' perception of how the system is structured and how their decisions might influence the system. Mental models reflect decision-makers' past observations, experiences, assumptions, and values. Forrester emphasized the importance of these 'mental databases' as the key information source, outweighing written or numerical databases (Forrester, 1992). CLDs can serve as an effective way to elicit and represent mental models, as they are not directly observed (Richardson et al., 1994). The explicit representation of mental models can lead to a rigorous examination and modification of mental models for enhanced decision-making.

Build a shared mental model. Stakeholders often hold different mental models of the system. In group decision-making, aligning stakeholder mental models can reduce conflict or disagreement. In this case, CLDs serve as an effective 'boundary object' (Star and Griesemer, 1989), a tangible representation of the system that can be modified by all those participating in the group decision-making process (Black and Andersen, 2012).

Visualize dynamic complexity. Policy problems are often embedded in a system with dynamic complexities characterized by feedback processes, non-linear relationships and time delays (Sterman, 2000). Dynamic complexity challenges our cognitive capacity and hinders our learning process (Sterman, 1994). Visual representation in CLDs helps stakeholders to see the nature of the dynamic complexities in the system and, in so doing, better understand their policy implications.

Articulate dynamic hypotheses. Analysts studying a system form a hypothesis of the structure that gives rise to the dynamic behaviour of interest. The CLD presents a visual representation of this *dynamic hypothesis*.

Methodology

> *In general, sufficient information exists in the descriptive knowledge possessed by the [stakeholders] to serve the model builder in [their] initial efforts ... Searching questions, asked at points throughout the [system] under study, by one skilled in knowing what is critical in system dynamics, can divulge far more useful information than is apt to exist in recorded data.*
>
> Jay W. Forrester (1961: 58–9)

Building blocks of CLDs

The following are conventions that have been established for the drawing of CLDs.[4] The basic structure involves a causal relationship between two elements in the system that is represented by two nodes (i.e. variables), and an arrow connecting those two nodes with a polarity.

Variables identify important system components. They are defined as a noun or noun phrase, and they must have a sense of direction in terms of quantity or degree (Sterman, 2000). For example, *patient health* can either improve or degrade, and *healthcare costs* can either increase or decrease. In contrast, *patient experience* does not have a similar sense of direction; *patient satisfaction* or *patient trust* would be clearer and permit the variables to distinguish among experiences. If variables are named incorrectly, it can lead to an error in specifying whether a causal link is positive or negative, which determines whether or not the loop is reinforcing or balancing, potentially leading to incorrect interpretation (Ford, 2009).

When defining a variable, it is helpful to make a distinction between actual system components and the perception of system components by different stakeholders. For example, *quality of care* would be different from *patient's perception of quality of care*. It is also helpful to differentiate target/desired states and current states. For example, *target quality of care* would be different from *actual quality of care*.

Arrows link variables to indicate a cause–effect relationship. All variables are connected to at least one other variable using arrows. Each arrow has a *polarity* to show the nature of the causal relationship (denoted by a plus or minus sign at the arrow head). Polarity between two variables describes the relationship between changes in those variables where all other variables in the system are held constant. A positive polarity in A to B indicates change in the same direction. In other words, an increase in A results in an increase in B. Likewise, a decrease in A results in a decrease in B. A negative polarity in C to D indicates change in opposite directions: an increase in C results in a decrease in D. Likewise, a decrease in C results in an increase in D (Ford, 2009; Sterman, 2000).

When there are *time delays* between cause and effect variables, they are indicated by a double line crossing the stem of the arrow (see Figure 6.1).

Causal relationships among variables can create two types of *feedback loops*. The first, *reinforcing (R) loops*, describe a positive feedback process where the loop generates a self-reinforcing behaviour – as in a vicious cycle or a virtuous cycle. Reinforcing loops can explain exponential growth or self-reinforcing decline behaviours in systems where the rate of change increases over time (see Figure 6.2).

Balancing (B) loops, in contrast, describe a negative feedback process where the loop generates a self-correcting, goal-seeking or stabilizing behaviour. The rate of change decreases over time, and the loop brings the system to equilibrium. The balancing loops can explain a system reaching a desired state or a system resisting a policy intervention that is attempting to move it away from the equilibrium state (see Figure 6.3).

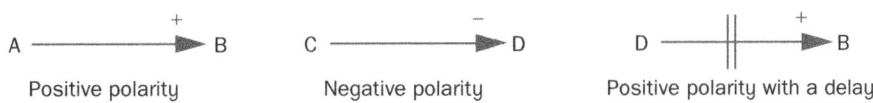

Figure 6.1 Polarity and a delay

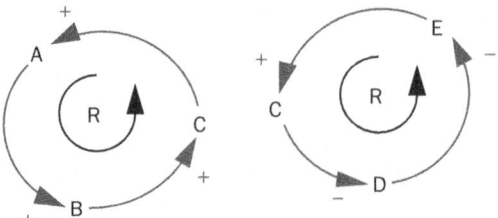

Figure 6.2 Two reinforcing feedback loops

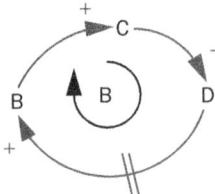

Figure 6.3 Balancing feedback loop with a delay

The **loop polarity** is determined by counting the number of negative links in the loop. If the number of negative links is even, then the loop is reinforcing; these reinforcing loops are also referred to as *positive loops*. In contrast, if the number of negative links is odd, then the loop is balancing; and balancing loops are also referred to as *negative loops*.

In CLDs, feedback loops are labelled with **loop identifiers** as follows:

1. Draw a circular arrow indicating which way – clockwise or counter-clockwise – to go about reading the loop.
2. Inside the circular arrow, use '**R**' or '+' for a reinforcing loop and '**B**' or '–' for a balancing loop.
3. Add a phrase to name the loop.
4. If there are many loops, number the loops to improve the clarity of the diagram (see Figure 6.4) (Sterman, 2000).

It is important to note that CLDs can also be drawn to include stocks and flows. In public health, stocks and flows are commonly used to show the population state transitions (for definitions, see Chapter 10). A classic example is when a population is exposed to an infectious disease: a portion of the members of the population move (or flow) from the state of being 'susceptible' to being 'infected' and then to being 'recovered' (commonly referred to the 'SIR model'). But what causes people to become infected? What causes them to recover? Is there feedback between those already infected and exposing new members of the population? These pieces can be drawn as causal links. We have a series of stocks and flows or an 'ageing chain' where a portion of the people in a stock will flow into the next stock and a portion of those then flow into the next one and so on. This chain acts as a backbone for drawing the causal links facilitating or acting as barriers to progression down the chain.

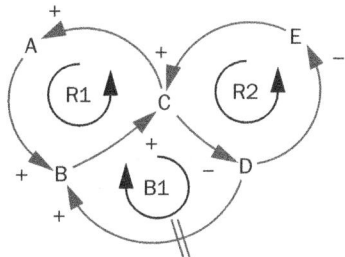

Figure 6.4 A causal loop diagram

Coloured text and/or arrows as well as bold font can be used to highlight individual feedback loops. Arrows are curved and have solid stems unless otherwise noted (e.g. using dashed stem for relationships that are less certain). Important feedback loops should be drawn in circles or ovals to increase the readability. System dynamics modelling software (for example, Vensim and Powersim) can be used to draw CLDs.

The following should not appear in a CLD (Sterman, 2000; Yearworth and White, 2013):

- An arrow linking a variable to itself.
- A link with ambiguous polarity.
- A correlation. The CLD is to represent causation not correlation.
- A variable having more than three to four arrows going into it. In this case, spell out the intermediate variables to clarify the pathways. Otherwise, it will indicate that we do not really understand the causal mechanisms/pathways.
- Outlines around a variable. Especially square outlines, as they imply a 'stock' in the system[5] and carry an important implication discussed in Chapter 10.

Successful CLD development depends much on the art of modelling. Here are some useful tips (Sterman, 2000):

- Choose the right level of aggregation; for example, making the intermediate step more explicit or removing it depending on what is most clear for the audience.
- Make the goals of negative loops explicit.
- Reduce cluttering – split the CLD into several smaller CLDs so as to retain clarity and build up the story piece by piece.
- Iterate. Redraw the CLD until crossing arrows are minimized and the most important loops are easy to read.

Process for developing CLDs

In this section, we present a typical process used in developing a rigorous CLD. System dynamics is concerned with problematic behaviour over time and CLDs help to focus on the endogenous causes of this behaviour. Although one can develop a CLD based upon one's own thinking about a problem, building more rigorous CLDs involves recognizing relevant theories and using a multi-method, and hopefully multi-stakeholder approach.

Problem definition

Before developing a CLD, a problem statement is drafted that focuses subsequent CLD development such that a system boundary is made clear – including the necessary elements, the appropriate timescale, and sufficient details so that the problem under investigation is endogenously produced. Without this, a CLD can quickly become a messy diagram that can overwhelm policy-makers without adding value.

Sterman (2000) recommends the following questions be considered:

- What is the key problem?
- What are the relevant concepts and system variables?
- What is the time horizon?
- What happened to these system variables in the past? What is likely to happen to these variables in the future? (Reference Modes)
- What explains the described behaviours of the selected variables? (Dynamic Hypotheses)

Furthermore, Sweeney and Meadows (2010) recommend thinking purposefully when choosing the timeframe for the analysis. They also emphasize the importance of defining the CLD context and audience. This is because defining these elements differently will likely change the goals, the paradigm, as well as the variables to include or exclude (or *model boundary*) and thus produce a different result. For example, consider the functioning of a primary care clinic. If we are interested in patient flow within the clinic, we might focus on a single workday. If we are interested in how that clinic's care management programme impacts readmissions, we might focus on a week to a month. But if we are interested in patient behaviour change efforts, we might focus on three months to a year. The audience may be the clinic staff, but it could also be the clinic management or the regional policy-maker.

To develop one's problem statement, a review of the relevant literature should first be conducted to better understand the problem under investigation. Also, discussions with stakeholders can provide added direction and clarity. In order to represent stakeholder perspectives in a balanced manner, a systematic method can be used to select key participants (Elias et al., 2001; Mitchell et al., 1997).

Data collection and generation of CLDs

CLDs can be generated from data that have previously been collected, or they can be generated in real time with stakeholders. Researchers may collect new data for the purpose of generating CLDs, or they may use existing data collected for a different purpose. The data may be generated at the individual level (then the results are merged by the researcher), or at the group level. Researchers can use archival data, direct observations, structured, semi-structured or unstructured interviews, or facilitated group discussions. Researchers may also use print media (e.g. books, newspaper articles) as well as social media (e.g. blogs, Facebook posts). The goal is to gather data that describe the underlying causes of the problem under investigation such that a dynamic hypothesis can be made.

Where possible, direct interviewing with CLDs in mind will conserve resources. Nevertheless, many situations will benefit from or even require the use of secondary data, such as in cases of problems at highly aggregated levels, where it may be difficult to recruit appropriate stakeholders with sufficiently broad experience.

While the CLD allows flexibility in its data source and the format for collection, a temporal and spatial distance between data source and researcher may introduce biases into the CLD (e.g. variables considered important for one researcher might be ignored by another). In that case, systematic coding[6] can be adopted to treat the data in a consistent manner. One method for coding text data for CLD development is Kim and Andersen's (2012) *purposive text analysis*. They used the verbatim transcripts from a macro-economic policy-making meeting to elicit a CLD. The CLD represents the policy-makers' mental models communicated and shared during the meeting leading to their collective decision. When the researcher cannot verify the CLD with the original stakeholder, systematic coding and documenting allows the researcher to leave a trace of data–CLD linkage and, where feasible, creates an opportunity for the CLD to be examined by others.

The coding procedure can be summarized as follows:

1. Define the problem of focus.
2. Select data segments within the problem boundary. Each data segment consists of one argument and its supporting rationales.
3. From each data segment, identify the cause variable, effect variable, and the polarity of the relationship.
4. Represent each causal relationship in a simple words-and-arrow diagram.
5. Collect and merge the words-and-arrow diagrams into a collective CLD. In doing so, collapse similar variables using a common variable name.
6. Assign unique identifiers to data segments and CLD variables and causal links. As the coding progresses, document the data segments where each CLD variable or causal link is elicited.

By drafting CLDs in real time with stakeholders, we can build confidence in the CLDs without requiring a systematic coding process (e.g. purposive text analysis). When a CLD is used during a semi-structured interview, this implies generating and reviewing the CLD with the respondent (see the cognitive mapping for CLDs example below). When facilitating a group, practices developed by system dynamics researchers for *group model building* (GMB; Richardson and Andersen, 1995; Vennix, 1996) might be useful for building confidence in CLDs. This process generates a greater sense of ownership and commitment to the understanding and recommendations generated. Facilitation is used to aid the group in developing a shared understanding of the system and its leverage points. GMB is typically used in interventions where a simulation model is the final result, yet many of the practices will be useful to developing CLDs. There is an open forum for sharing GMB facilitation best practices (see Hovmand et al., 2013).

The specific CLD development method chosen depends upon the data and resources available, skills/comfort level of the researcher, as well as the purpose of the CLDs. For example, purposive text analysis utilizes existing text (or it can be used to analyse newly generated text), while one-on-one and GMB requires interaction with stakeholders. Purposive text analysis relies heavily upon qualitative analysis skills, whereas one-on-one and GMB relies heavily upon interviewing and group facilitation skills. One-on-one model building may better capture minority thoughts, whereas GMB relies upon skilfully crafted groups and real-time facilitation to protect the minority's comfort level with sharing thoughts. If there is a need to build a shared understanding of the system, then the GMB process will lead to a better result.

The key to identifying dynamically important information is not in how the original source was generated, but in the skill of the researcher in recognizing when a system structure exhibits the signs of feedbacks, time delays, and non-linearities.

Verification and modification of CLDs

Throughout the analysis process, CLDs will grow new branches, some of which are grafted back in and some of which are pruned. The key is to iterate the CLD through this stage. While well-known quantitative tests of accuracy, such as the *P*-value, do not readily apply to CLDs, various tools have been developed to assess CLDs.

One method for assessing CLDs is the *disconfirmatory interview* (for specific examples, see Andersen et al., 2012). This method builds on previous methods to facilitate the conversation about the adequacy of the CLD developed. In disconfirmatory interviews, stakeholders are encouraged to question the CLD structure and point out flaws and biases in the CLD rather than confirming the existing CLD structure. Based on the interview responses, the CLD will be further modified.

Users of CLDs (or more broadly simulation models based on CLDs) recognize that all models are simplification of the real world (Sterman, 2002). Practical criteria for evaluating CLDs involve *usefulness* as opposed to *validity*. According to Forrester, validation is a process of establishing confidence in the soundness and usefulness of a model (Forrester and Senge, 1996). Dynamic models should be useful, illuminating, convincing, and inspire confidence (Greenberger et al., 1976). In other words, does the CLD generate learning? Does it allow for a new level of understanding, hopefully allowing for better policy design and decision-making? Does it help stakeholders understand each other's point of view? Does it create *shared* understanding?

Applications

This section presents selected case studies of health systems work, each focusing on a distinct technique for developing CLDs (i.e. eliciting and verifying mental models). They represent examples of research at the health system level where CLDs played an integral role. Each case presents an adapted variant of the methods presented above, based on the needs, resources, and knowledge of the researchers at the time. As such, the methods used in each case have relative strengths and limitations and are not to be considered the gold standard. That said, each provides a concrete example of the usefulness and added value CLDs permit.

Purposive text analysis: redesigning primary care in Utah, USA

The patient-centred medical home (PCMH) is a service delivery model that aims to improve the efficiency and effectiveness of primary care. The objective of the research carried out by the team in Utah was to build understanding of how healthcare delivery organizations could achieve and sustain PCMH implementation (Tomoaia-Cotisel et al., 2013). Funding was provided to a set of similar projects across the US beginning in 2010.

A core aspect of PCMH implementation at the study site in Utah was task shifting: the delegation of patient care tasks previously performed by clinicians to supervised medical assistants (MAs). The choice of whether or not to implement is in the hands of the clinician (a physician assistant, nurse-practitioner or physician).

Data collection began with *semi-structured interviews* conducted with over 80 individuals in the health system. Interview questions focused on the structure of the programme being implemented, how it was being implemented and the results of that implementation. During the interviews, many respondents commented that achieving PCMH was difficult because it involves successfully balancing a complex set of trade-offs. As the mixed-methods analysis proceeded, it became apparent that to understand PCMH and to generalize lessons learned, more attention had to be paid to the tensions within the primary care system.

Purposive sampling was used to identify one medical assistant and one clinician from each care team, from five primary care clinics deemed to be representative (in their context) of the larger organization. *Purposive text analysis* proceeded as follows: 20 interview transcripts were coded to identify portions where structural relationships were discussed, focusing on statements describing a cause-and-effect relationship.

Simultaneous with coding in computer-aided qualitative data analysis software (CAQDAS), respondent mental models were visualized in system dynamics modelling software. For this, the coded causal relationships were then translated into a single words-and-arrow diagram for each individual. These diagrams were then pruned – retaining only linkages involving delays or loops of three or more links (following Yearworth et al., 2013). Next, variable names were standardized across all models. Individual CLDs were then merged into team CLDs. Team CLDs were then merged into clinic CLDs. At this point, the CLD structure was essentially the same, regardless of clinic, indicating that conceptual saturation was being reached.

The final stage was to *verify the CLD* with various key informants. These discussions were held to review the preliminary CLD. Participants included expert researchers, clinicians, managers, and health systems experts from both inside and outside the study site.

This example will illustrate how purposive text analysis was used. One MA (MA03) describes task shifting within her team as follows. (*Note*: for PCMH at the study site, task shifting requires the MA to stay in the examination room during the visit.)

So we pretty much do it all except for staying in the room with the clinician. And [when] this is brought out, Dr XXX will start doing it again for a week or so and then she'll stop again. I think that we get so involved with what we're doing . . . we work shorthanded a lot like we've been very shorthanded today. [1] . . . [Plus, the message we get so often is: you're] not seeing enough patients, not seeing enough patients, not seeing enough patients [2], and, you know, we're not making the money. [3] . . . and unfortunately sometimes I think that its pushed too much. And the quality of care is not there. [4]

Figure 6.5 shows how this segment was subdivided into four causality arguments, and a corresponding words-and-arrow diagram, in the words of medical assistant MA03.

Bringing individual mental models together provided a fuller picture of the mechanisms and interdependencies giving rise to the system tensions faced by clinicians, including a clear description of the difficult choices that implementing PCMH requires, trading off short-term costs for long-term benefits.

Cognitive mapping for CLDs: commuting and wellbeing in Auckland, New Zealand

Commuting has complex linkages with health and the social determinants of health. This complexity, together with feedbacks inside the transportation system, makes it difficult to determine appropriate policies. The objective of the research team was to better understand the links between commuting and wellbeing in Auckland, with a focus on social equity, so that policies might be identified that optimize the wellbeing outcomes of regional transport policies (Macmillan et al., 2014, 2016).

Main argument: we don't bring the MA into the room because it slows us down when we are already short-staffed					
		Argument 1	**Argument 2**	**Argument 3**	**Argument 4**
Quotation	M03	*I think that we get so involved with what we're doing . . . we work shorthanded a lot . . .*	*[Plus, the message we get so often is: you're] not seeing enough patients . . .*	*and, you know, we're not making the money.*	*. . . and, unfortunately, sometimes I think that it's pushed too much. And the quality of care is not there.*
Causal structure	Causal variable	working short-handed	bringing the MA into the room	seeing enough patients	pushing a focus on seeing enough patients
	Effect variable	bringing the MA into the room	seeing enough patients	making the money	quality of care
	Relationship type	Negative	Negative	Positive	Negative

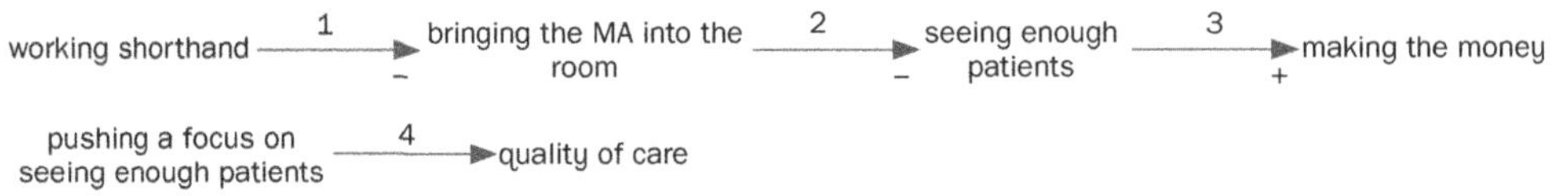

working shorthand —1→ bringing the MA into the room —2→ seeing enough patients —3→ making the money

pushing a focus on seeing enough patients —4→ quality of care

Figure 6.5 The purposive text analysis of MA03, arguments 1 to 4

Sixteen participants representing the diversity of stakeholders involved (in this case, those designing, influencing or affected by transportation policy in Auckland) were identified via purposive sampling. *Semi-structured interviews* focused on the co-creation of cognitive maps. In the introduction to the interview, the interviewer explained that the process would result in iterative knowledge generation that would inform policy as well as empower participants to act.

All interviews began with the following question:

Can you tell me a bit about your views on the links between commuting (how people get to work) and the wellbeing of your community?

Participants were asked to consider the regional-level impacts rather than those solely of their neighbourhood or locality. Interviews were not transcribed since their purpose was the creation of the cognitive map. Both the researcher and the participant had a pencil and eraser, and drew the map together.

As themes emerged, follow-up questions were structured as follows:

Let's talk a bit more about the causes of this – why/how does this happen?

Let's talk a bit more about the consequences of this – what happens because of this?

During each interview, several *cognitive maps* were generated using pencil, eraser, and paper. Depending on the themes that arose, the maps could be expanded over the course of the interview. Maps were refined through discussion and reviewed one last time until no further themes or changes emerged. Figure 6.6 presents a scanned cognitive map from one of the interviews. It shows, on the left, reasons people choose to take public transportation and, on the right, the benefits.

With these cognitive maps in hand, a participatory workshop was held, where participants organized variables and relationships from the cognitive maps into a single set of interlinked themes. The research team then pruned these relationships so that only those in feedback loops remained. A literature review then served to test whether relationships identified in the literature either refuted or supported the set of feedback loops that emerged (now, constituting a dynamic causal theory for the problem studied). The resulting *preliminary CLD* was then verified and further revised in a second workshop with stakeholders, as well as an additional set of individual and organization-specific meetings with relevant stakeholders, resulting in a refined final CLD.

The *final CLD* presents a causal theory for the observed trends in and impacts of commuting in Auckland. A portion of this CLD was the focus of follow-on work specifically to explore commuting by bicycle. The CLD provided a grounded, dynamic theoretical framework from which to explore this specific problem in more detail. The loops involving cycling were expanded via further stakeholder engagement and quantitative simulation modelling. The resulting model informed Auckland's regional planning that was aiming to promote cycling for transportation in the midst of exponential growth in car ownership and use. Specifically, it evaluated Auckland's existing plan (developing a regional cycling

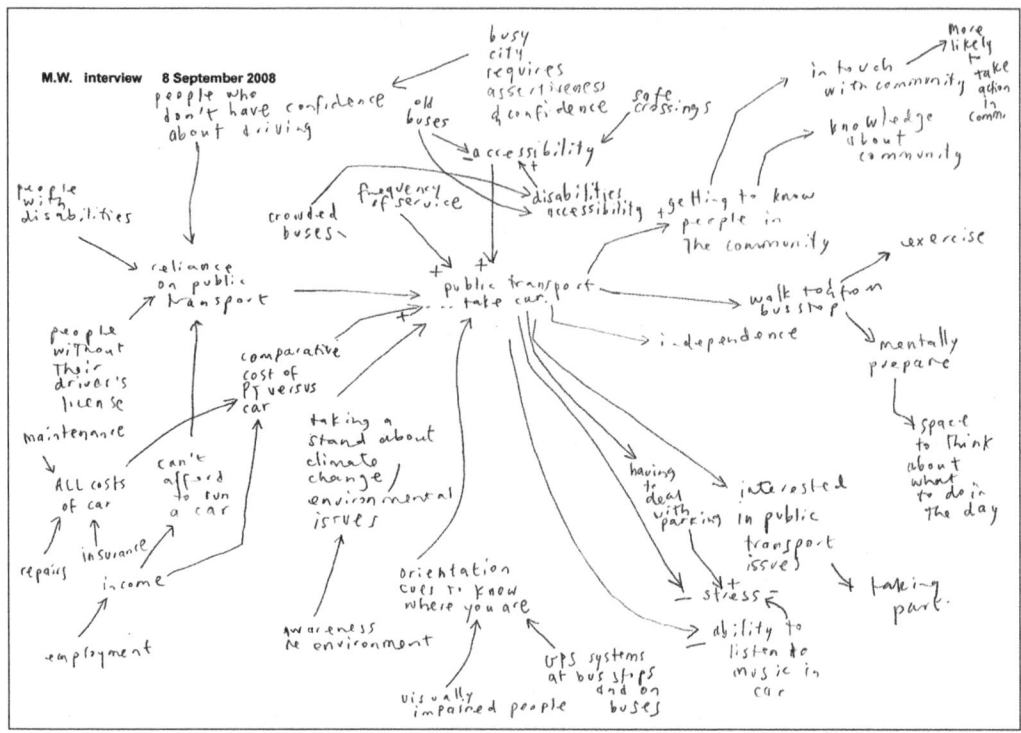

Figure 6.6 Cognitive map drawn during a stakeholder interview

network), finding that it was not sufficient to reverse the cycling injury trend that limits the adoption of cycling. It then also provided insight into potential plans for a deeper redesign of the road infrastructure to accommodate safe cycling across the city.

Expert CLD creation: neonatal mortality in Kampala District, Uganda

The first four weeks of life, known as the neonatal period, are when children are at their most vulnerable. The objective of this research was to build understanding of the persistently high rate of neonatal mortality in Uganda, even after implementation of best practice interventions. The research was funded through grants from the Canadian government and the World Health Organization (Semwanga et al., 2014, 2016).

Research on the subject of neonatal mortality has identified many associated variables, resulting in immensely detailed complexity around this problem. Stakeholders may perceive additional causes. This complexity complicates the process of identifying high-leverage interventions, making the problem more intractable for policy-makers in Uganda.

CLD development focused on documenting this intricate complexity, and began by identifying a set of variables and thematic categories using a *literature review. Semi-structured interviews* were conducted to explore stakeholders' (e.g. mothers, health workers, community leaders, policy-makers) experiences and attitudes with respect to the entire episode of care from pre-conception to postnatal care as well as experiences of women seeking to be future mothers.

The research team then proceeded to reduce the set of variables to a list of potentially important variables using *thematic coding* and then, in the deductive tradition of expert modelling, proceeded to develop CLDs in *brainstorming sessions*. Finally, a *survey* was created to verify the CLD. This survey was sent to expert stakeholders: local and international researchers as well as implementers. The survey was sent in the form of a spreadsheet, where respondents gave open-ended responses to the questions below for pieces of the CLD:

1. Do all the variables stated in this (CLD) exist?
2. Do the relationships between these variables exist?
3. Are there any significant causal factors missing? If so, list them.
4. Are the directions of the links right or (do) they need to be reversed (implying that the effect is the cause).
5. What other effects could be observed as a result of this cause?

The results of this survey demonstrated that the CLD's intended audience found it to have potential as a useful tool for dialogue and decision-making, building confidence in this conceptual model. When new questions were raised and/or modifications were suggested, the study team considered them. Those deemed to merit inclusion were addressed using previously collected literature and interview data.

The resulting CLD permitted qualitative exploration of potential high-leverage interventions that included: promoting girls' education, ensuring sufficient socio-economic resources to enable mothers to access the health system, and improving procedures (e.g. supervision and auditing) and resources (e.g. staffing, logistics, and medicines) at health facilities. This CLD then became the basis for the development of a quantitative policy simulation model. Figure 6.7 presents a CLD showing two loops adapted from the authors' final CLD. It illustrates the tension

Figure 6.7 Causal loop diagram
Source: adapted from Semwanga et al. (2014).

between the need to build awareness of services in the community to generate new demand while also meeting the existing demand for those services with sufficient resources.

Discussion

A decade ago, Green (2006) made four requests of systems science on behalf of the field of public health:

1. Help unravel/re-ravel the many mediating and moderating variables that impact the successful scale up of an intervention – taking into account the multiple levels, the recursive relationships between biology, behaviour, and the environment.

   Stakeholders in health systems and public health regularly deal with issues of great complexity. While they can easily describe two variables that they struggle to keep in balance, they have a difficult time describing the causal mechanisms linking them. The cases described herein demonstrate how CLDs can merge stakeholder mental models to uncover the causal mechanisms between mediating variables.

2. Help move beyond linear analyses, beyond interaction effects in ANOVA and regression, to cause and effect and feedback loops.

   CLDs move beyond linear analyses by drawing causal loops that identify the feedbacks, delays, stocks, and flows underlying non-linearities.

3. Help move beyond testing interventions in artificial environments controlling for exogenous variables, to an inductive approach assessing within the practice setting.

   CLDs facilitate an inductive approach where understanding is gained within the real environment, based on the experience of practitioners. This allows for expanding the definition of a problem to make previously exogenous variables endogenous. However, testing requires a quantitative model, as described in Chapter 10.

4. Is chaos theory so central to systems science that public health would have to come to grips with adopting it?

   CLDs describe problems where feedback loops interact. Some of these problems exhibit the repeating cycles of chaotic systems arising from balancing feedback loops accompanied with time delays (see Sterman, 2000). Many such feedback loops have been modelled in the cases described here without the expected conceptual difficulty.

Much good CLD and simulation work has been done in the broad domain of health. A recent virtual issue of the *System Dynamics Review* (Homer et al., 2015a) includes the re-issue of papers considered to be the best applications of system dynamics in health, including a bibliography that is intended to catalogue applications of system dynamics in the health domain (Homer et al., 2015b).

Readers interested in applying these methods should include these documents in their literature reviews and in seeking out mentors.

Conclusion

Pencil and paper, chalk and chalk board: these are the minimum necessary resources that a person needs to begin developing CLDs. With practice, you will get beyond just linking causal relationships and learn to recognize the time delays, accumulations, and feedback loops in others' verbal descriptions of problems. We hope that the tools we have presented in this chapter will empower people[7] to develop CLDs with firm grounding in the practical experience of stakeholders, which communicate both general truths and specific counter-intuitive insights, and which policy-makers can use and understand.

Chapter 10 presents methods for testing the structural hypotheses developed here by translating CLDs into mathematically formulated computer simulation models. By building a model that can quantitatively reproduce the problem behaviour (i.e. the reference mode), it is possible to test if the hypothesized system structure would generate the observed system behaviour and, through an iterative process, to refine the hypothesis and come to better understand the system. Formal simulation models can answer some of the questions that remain with CLDs – for example:

- If there are multiple feedback loops in the system, under what conditions does one loop dominate another?
- If there are multiple leverage points, which one or which combination has the greatest impact?

However, formal simulation modelling can be costly in terms of the time, resources and effort required. Stakeholders' goals and objectives must be considered when deciding CLDs' role as a stand-alone method or as a conceptualization tool for formal modelling.

Notes

1 As cited in Carus (1890: 14).
2 CLDs are one tool emerging from system dynamics founder Jay Wright Forrester's prolific efforts to transform problem formulation. In his words: 'We need new attitudes toward the construction and use of models of social systems . . . [We should] make better use of our vast store of descriptive information . . . Both physical and psychological factors can be included. [Techniques] no longer limit the systems that can be studied . . . Such models will communicate easily with [stakeholders] because they arise from the same sources and in the same terminology as [their] experience' (Forrester, 1961: 53–5, 59).
3 See the 'Six shapes to describe dynamic behavior' in Ford (2009) [http://public.wsu.edu/~forda/Ch%201.pdf].
4 Free software is available for educational purposes: Vensim PLE.
5 System dynamics is the proper term for what is discussed in this chapter. It is different from systems dynamics, systems dynamic, and dynamical systems. While each of these methods looks at problems of system dynamics, CLDs are only used in system dynamics modelling.

6 Systematic coding, in this context, is defined as qualitative coding of text data where text is interpreted for causal structures. This goal of eliciting causal structures differs from thematic coding.
7 Additional resources include: System Dynamics Society conferences and training events (see systemdynamics.org) as well as mentoring (see Richardson et al., 2015).

Acknowledgements

We would like to thank Luis Luna-Reyes, PhD for his review of the chapter and comments and Alexandra Macmillan, PhD for her assistance in developing one of the cases.

References

Adam, T. and de Savigny, D. (2012) Systems thinking for strengthening health systems in LMICs: need for a paradigm shift, *Health Policy and Planning*, 27 (suppl. 4): iv1–iv3.
Andersen, D.L., Luna-Reyes, L.F., Diker, V.G., Black, L., Rich, E. and Andersen, D.F. (2012) The disconfirmatory interview as a strategy for the assessment of system dynamics models, *System Dynamics Review*, 28 (3): 255–75.
Axelrod, R. (1976) *Structure of Decision: The cognitive maps of political elites*. Princeton, NJ: Princeton University Press (reprinted 2015).
Black, L.J. and Andersen, D.F. (2012) Using visual representations as boundary objects to resolve conflict in collaborative model-building approaches, *Systems Research and Behavioral Science*, 29 (2): 194–208.
Bougon, M.G. (1992) Congregate cognitive maps: a unified dynamic theory of organization and strategy, *Journal of Management Studies*, 29 (3): 369–87.
Carus, P. (1890) *The Ethical Problem: Three lectures delivered before the Society for Ethical Culture of Chicago in June 1890*. Chicago, IL: Open Court Publishing Company.
de Savigny, D. and Adam, T. (2009) *Systems Thinking for Health Systems Strengthening*. Geneva: World Health Organization.
Eden, C., Ackermann, F. and Cropper, S. (1992) The analysis of cause maps, *Journal of Management Studies*, 29 (3): 309–24.
Elias, A.A., Cavana, R.Y. and Jackson, L.S. (2001) Stakeholder analysis to enrich the systems thinking and modelling methodology, in J.H. Hines, V.G. Diker, R.S. Langer and J.I. Rowe (eds.) *Proceedings of the 19th International Conference of the System Dynamics Society* (pp. 52–3). Albany, NY: The Systems Dynamic Society.
Ford, F.A. (2009) *Modeling the Environment*, 2nd edn. Washington, DC: Island Press.
Forrester, J.W. (1961) Models, in *Industrial Dynamics*. Cambridge, MA: MIT Press.
Forrester, J.W. (1992) Policies, decisions and information sources for modeling, *European Journal of Operational Research*, 59 (1): 42–63.
Forrester, J.W. and Senge, P. M. (1996) Tests for building confidence in system dynamics models, *Modelling for Management: Simulation in Support of Systems Thinking*, 2: 414–34.
Green, L.W. (2006) Public health asks of systems science: to advance our evidence-based practice, can you help us get more practice-based evidence?, *American Journal of Public Health*, 96 (3): 406–9.
Greenberger, M., Crenson, M.A. and Crissey, B.L. (1976) *Models in the Policy Process: Public decision making in the computer era*. New York: Russell Sage Foundation.
Homer, J., Hirsch, G. and Tomoaia-Cotisel, A. (eds.) (2015a) System dynamics applications to health and health care, *System Dynamics Review* [available at: http://onlinelibrary.wiley.com/journal/10.1002/(ISSN)1099-1727/homepage/VirtualIssuesPage.html].

Homer, J., Hirsch, G. and Tomoaia-Cotisel, A. (eds.) (2015b) System dynamics applications to health and health care: bibliography, *System Dynamics Review* [available at: http://onlinelibrary.wiley.com/journal/10.1002/(ISSN)1099-1727/homepage/SDR-Virtual-Health-BIBLIOGRAPHY.pdf].

Hovmand, P.S., Rouwette, E.A.J.A., Andersen, D.F., Richardson, G.P. and Kraus, A. (2013) *Scriptapedia Version 4.0.6*. Paper presented at the 31st International Conference of the System Dynamics Society, Cambridge, MA [available at: www.systemdynamics.org/conferences/2013/proceed/papers/P1405.pdf; accessed July 2017].

Kim, H. and Andersen, D.F. (2012) Building confidence in causal maps generated from purposive text data: mapping transcripts of the Federal Reserve, *System Dynamics Review*, 28 (4): 311–28.

Macmillan, A., Connor, J., Witten, K., Kearns, R., Rees, D. and Woodward, A. (2014) The societal costs and benefits of commuter bicycling: simulating the effects of specific policies using system dynamics modeling, *Environmental Health Perspectives*, 122 (4): 335–44.

Macmillan, A., Davies, M., Shrubsole, C., Luxford, N., May, N., Chiu, L.F. et al. (2016) Integrated decision-making about housing, energy and wellbeing: a qualitative system dynamics model, *Environmental Health*, 15 (suppl. 1): S37.

Mitchell, R.K., Agle, B.R. and Wood, D.J. (1997) Toward a theory of stakeholder identification and salience: defining the principle of who and what really counts, *Academy of Management Review*, 22 (4): 853–86.

Richardson, G.P. and Andersen, D.F. (1995) Teamwork in group model building, *System Dynamics Review*, 11 (2): 113–37.

Richardson, G.P., Andersen, D.F., Maxwell, T.A. and Stewart, T.R. (1994) Foundations of mental model research, in C. Monaghan and E. Wolstenholme (eds.) *Proceedings of the 1994 International System Dynamics Conference: Problem-solving methodologies* (pp. 181–92). Stirling: University of Stirling.

Richardson, G.P., Black, L.J., Deegan, M., Ghaffarzadegan, N., Greer, D., Kim, H. et al. (2015) Reflections on peer mentoring for ongoing professional development in system dynamics, *System Dynamics Review*, 31 (3): 173–81.

Semwanga, A.R., Nakubulwa, S., Nakakeeto-Kijjambu, M. and Adam, T. (2014) Advancing the application of systems thinking in health: understanding the dynamics of neonatal mortality in Uganda, *Health Research Policy and Systems*, 12 (1): 36 [doi: 10.1186/1478-4505-12-36].

Semwanga, A.R., Nakubulwa, S. and Adam, T. (2016) Applying a system dynamics modelling approach to explore policy options for improving neonatal health in Uganda, *Health Research Policy and Systems*, 14 (1): 35 [doi: 10.1186/s12961-016-0101-8].

Star, S.L. and Griesemer, J.R. (1989) Institutional ecology, translations and boundary objects: amateurs and professionals in Berkeley's Museum of Vertebrate Zoology, 1907–39, *Social Studies of Sience*, 19 (3): 387–420.

Sterman, J.D. (1994) Learning in and about complex systems, *System Dynamics Review*, 10 (2/3): 291–330.

Sterman, J.D. (2000) *Business Dynamics: Systems thinking and modeling for a complex world*. New York: Irwin/McGraw-Hill.

Sterman, J.D. (2002) All models are wrong: reflections on becoming a systems scientist, *System Dynamics Review*, 18 (4): 501–31.

Sterman, J.D. (2006) Learning from evidence in a complex world, *American Journal of Public Health*, 96 (3): 505–14.

Sweeney, L.B. and Meadows, D. (2010) *The Systems Thinking Playbook: Exercises to stretch and build learning and systems thinking capabilities*. White River Junction, VT: Chelsea Green Publishing.

Tomoaia-Cotisel, A., Scammon, D.L., Day, J., Day, R., Kim, J., Waitzman, N. et al. (2013) Connecting the dots and merging meaning: using mixed methods to study primary care delivery transformation, *Health Services Research*, 48 (6 pt. 2): 2181–2207.

Vennix, J.A.M. (1996) *Group Model Building: Facilitating team learning using system dynamics.* Chichester: Wiley.

Yearworth, M. and White, L. (2013) The uses of qualitative data in multimethodology: developing causal loop diagrams during the coding process, *European Journal of Operational Research,* 231 (1): 151–61.

7 | Network analysis: a tool for understanding social network behaviour of a system

Karl Blanchet and Jessica Shearer

Introduction

Social network theory and empirical research have demonstrated the influence of interpersonal relationships, and the larger structure they create, on decision-making and a range of other individual- and system-level behaviours and outcomes. Individuals are connected to other individuals, forming a social network, which Milgram (1967) called a 'small world'. Asch (1953) showed that individuals' decisions in an unpredictable world are often based on peers' opinions and actions. Interactions and collaboration between stakeholders depend on various social factors such as trust, conflict resolution, and knowledge integration (Folke et al., 2002). However, all these factors rely on the circulation of information within social networks (Bodin et al., 2006; Olsson et al., 2004). Studying information flow mechanisms between actors and within networks can help us to understand decision-making processes, as well as the social processes that influence the resilience of socio-ecological systems (including complex health systems).

Since the 1990s, with the development of mathematical models, social networks have gained prominence in the literature; however, social network theories have a long history in health care. The role of networks has become crucial in health care during the twenty-first century with the emergence of informational and technological innovations, and with recognition by health managers that hospitals are no longer the only site of healthcare delivery (Thompson et al., 2006). Healthcare providers have recognized the role of other actors – medical and non-medical, private and public – as well as the positive impact of multi-scale and multidisciplinary network-based initiatives (Thompson et al., 2006). In health systems research, networks have implicitly been at the heart of health systems (WHO, 2010). In their own definition of a health system, Kohn et al. (2000) made even more explicit how social networks play a crucial role. They saw a health system as a network of actors who aim to provide health care: 'In health care, a system can be an integrated delivery system, a centrally owned multihospital system, or a virtual system comprised of many different partners over a wide geographical area' (Kohn et al., 2000: 52).

Social networks and health care also have a common history. Social network theories were born in public health when, in 1934, after an epidemic in a New York school, Moreno tried to understand why the pandemic had spread so quickly among the pupils. Moreno (1934) was also the first to represent graphically the relationships between pupils and their social position. In order to model systems, social networks theorists applied mathematical and graphical techniques to illustrate and understand the complexity of human and organizational relationships. A social network is composed of a set of actors (for example, in the health sector, hospitals, doctors, and patients) often called 'nodes' and represented by spots (Borgatti and Foster, 2003; Degenne and Forsé, 1999; Marsden, 1990), which are interconnected by a set of 'ties' or relationships represented by straight lines (Batley and Larbi, 2004; Islam, 2007).

The field of social network analysis (SNA) provides an avenue for analysing and comparing formal and informal information flows in a system. Although SNA and health care have long been interconnected, SNA has rarely been applied to health systems research in low- and middle-income countries, which remains a nascent field of investigation. SNA has been proven, in other areas, to be helpful in understanding the nature of relations between actors within a system and how these relationships influence the structure of a system (Borgatti et al., 2009; Webb and Bodin, 2008).

Although applying SNA in systems research is not without its challenges, such as capturing the dynamics of systems and the effects of cross-scale events, innovative approaches have been introduced combining social network theories and other approaches, and can potentially generate new knowledge when applied to health systems (Cumming et al., 2010). In ecology, studies using SNA have focused on the relationships between the different species in a food web and how the disappearance of one species could have a major impact on other species connected in the same food web. Social network theories have also been very useful when studying interactions between ecological systems and communities, and one can imagine how these concepts could be applied to health systems (Moreno, 1934; Spillane, 2006). Scholars have found that there is a relationship between the structure of networks, the type of links between actors (i.e. bonding between actors of the system or bridging links with other systems), and the resilience of socio-ecological systems (Burt, 2003; Newman and Dale, 2005).

In social sciences, scholars have shown that network structure determines the level of cooperation between individuals (Nowak, 2006); in other words, individuals tend to collaborate more easily with their direct neighbours and people who have similar characteristics and interests, described as *homophily* (Lazarsfeld and Merton, 1978). SNA researchers have also shown that, while individuals are connected with a limited number of people, all people in the world are indirectly connected by a number of ties, which, on average, does not exceed 'six degrees' (Watts and Strogatz, 1998). SNA was used, for example, to show that social connections represent a social capital that provides power to find jobs (McGuire, 2000) or finalize business contracts (Uzzi and Gillespie, 2002). This high degree of connectivity between individuals and organizations has implications for the level of *interdependence* and *embeddedness* between networks (i.e. the fact that networks are interconnected and influence each other). Individuals connected through a social network tend to have similar beliefs and values (Kiesler and Cummings, 2002). Christakis and Fowler (2007) even showed that people who are obese tend to be connected with other obese people. Social networks theory has also helped researchers understand the management and diffusion of knowledge (Borgatti and Cross, 2003), group behaviour, group dynamics, and organizational structures or power relationships within an organization or a group (Wasserman and Faust, 1994). SNA has also been used to analyse the patterns of diffusion of innovations and, in particular, how the structure of a network or system determines the degree of adoption of innovations.

Research questions

Network dependencies

What is a network? A network is a system characterized by a set of relationships between different actors, and the structure formed by these relationships. Any theory, finding or research question related to networks is situated on at least one of these levels: individuals ('nodes'), relationships ('ties'), and the resulting network. Prior to embarking on SNA, the researcher should have a clear understanding of their research question, what the network is, and where the question is situated in terms of the network or networks. Table 7.1 lists hypothetical network research questions, nodes, and relationships that a health systems researcher might study.

At the node level, SNA can be a valuable tool to uncover the most influential players in a system, those best placed to access, disseminate or control information, or to explain node-level behaviour change (Borgatti and Foster, 2003; Riggan and Supovitz, 2008; Valente and Pumpuang, 2007). Centrality, reachability, and betweenness are the most well-known node-related properties of networks that can be easily computed on a complete network (Padgett and Ansell, 1993). Table 7.2 defines these and other network measures. Potential knowledge brokers – individuals who are positioned to exchange information or research evidence through a health system – can be identified by their betweenness centrality (Borgatti et al., 2009). The brokers in a health system help coordinate actors in times of crises or shocks and build bridges between different groups within the system (Thompson et al., 2006). Other actors essential to the diffusion of innovations, such as opinion leaders, champions or change agents, can be identified through the number of links they have with their peers or non-peer actors at different levels of the health system (Burt, 2003; Newman and Dale, 2005; Shearer, 2015).

A network exists as soon as two actors (a dyad of nodes) are connected by a relationship (i.e. a 'tie' or 'edge' in network-speak). This level is primarily concerned with the description and estimation or modelling of social processes and exchanges. Advances in theory and methods increasingly point to the importance of understanding the relationships between entities to better understand complex systems.

Relationships may be *non-directional* (e.g. collaborating) or *directional* (e.g. providing information), *binary* or *valued*. Some relationships make *reciprocity* possible. For example, if A likes B, it is possible that B also likes A. In other instances, reciprocity will be impossible, for example when the relationship is

Table 7.1 Examples of systems, their nodes, and relevant relationships

Example	Nodes	Ties	Ex. Network research question
1. Health system	Health facilities	Supervisory or referral relationships	What is the fastest way to disseminate new directives to first-level health facilities?
2. Community healthcare network	All health providers (public, private, traditional) in a given community	Referral relationships, discussion or information relationships	Which network or node-level factors explain referral patterns in one village's healthcare network? (tie-level outcome)
3. Vaccine supply chain network	Cold stores, rooms, and fridges	Supply 'push' and 'pull' relationships	What is the ideal structure of a vaccine supply chain network?
4. Learning network	Health providers, managers or decision-makers participating in a peer-learning intervention	Knowledge exchange, advice, and discussion relationships	How does a peer-learning intervention change the size, density, and centralization of a knowledge exchange network? (whole network outcome)
5. Policy-making network	Stakeholders involved in a policy decision-making process (e.g. policy-makers, technocrats, development partners, civil society representatives)	Communication, information exchange, and resource exchange relationships	Which network- or node-level factors explain an actor's use of evidence to inform a policy decision?

Table 7.2 Definition of key network measures

Characteristic	Measurement level	Measure
Degree centrality	Node	Degree centrality is the count of a node's ties (Freeman, 1984). Directed networks will have in-degree and out-degree centrality
Betweenness centrality	Node	Betweenness is a measure that indicates how much a node is located in the path between other actors (Freeman, 1977). Betweenness is an important indicator of ability to control the flow of information or resources
Closeness centrality	Node	Closeness centrality measures how near a node is to all other nodes in the network, either through direct ties or through ties that go through others. Closeness reflects how quickly an actor can reach all others in a network
Reachability	Node	Reachability defines the degree by which a node can be reached by other nodes. If a certain number are unreachable by some actors, it means that the network is fragmented. Reachability corresponds to the number of steps maximally needed to reach from one node to any other node in the network
Distance	Node	Distance measures the number of ties that separate two actors. If two nodes are directly connected, the distance is one. If these two nodes are separated by one node, the distance is two
Density	Network	Density is defined as the number of existing ties divided by the number of possible ties. Dense (cohesive) networks are more likely to resist change, exchange non-complex information or act collectively, whereas sparse networks may be more open to new information and actors, and thus innovation
Centralization	Network	Centralization is a measure of the dispersion of all nodes' centrality scores. Centralized networks act more efficiently under the control of one or a few focal actors, whereas decentralized networks are better at finding and exchanging new information and ideas
Transitivity	Edge	Transitivity is a social process where Actor A is more likely to form a relationship with Actor C if there is already a relationship between Actors A and B and Actors B and C. The result is a triangle between Actors A, B, and C
Reciprocity	Edge	Reciprocity exists when a directed relationship exists in both directions
Multiplexity	Edge	Multiplex edges refer to multiple different relationships layered between two given nodes

defined as 'manager of'. A network can be characterized by nodes having common characteristics (e.g. a network of healthcare workers) or illustrating the flows circulating between one node and another (e.g. circulation of information). The notion of *transitivity* can be applied to networks with at least three actors (a triad of nodes). Triangles exist more often than chance alone in many social settings and have a range of consequences on the functioning of networks and systems.

At the network level, network structure has been associated with functions and outcomes across a range of disciplines and topics. In health systems research, it is important to make sense of the system as a whole and thus SNA can help us to understand how health systems are structured and how structure relates to function or outcomes. Two properties in particular are used in SNA: cohesion and shape (Lebel et al., 2000). Cohesion defines the number of connections within the network and includes sub-properties such as density and fragmentation. Shape relates to the overall distribution of ties and distinguishes the core actors from the peripheral ones (Borgatti et al., 2009).

Finally, while many SNA research questions remain cross-sectional in nature, social networks are dynamic systems, and thus benefit when possible from measurements over time. Truly longitudinal studies of networks help to identify why relationships form and dissolve, and how information or diseases spread. Mapping a network before and after an event, intervention or change can help identify the impact of that change on the network. Network analysis often helps to identify and quantify the unintended consequences of complex interventions.

Applying these constructs to a health system (see examples in Table 7.1), one imagines a multitude of potential networks to explore, including but not limited to networks of patients and their carers, networks of providers (different by cadre, speciality, public or private), managers, equipment and drugs companies, and funders (e.g. government, insurance companies, international donors). Different networks will be important for different health systems functions, including policy-making and governance, financing, planning, management, and evaluation. The composition and structure of these networks are changing all the time because the actors of the sub-systems are moving in and out of the network (e.g. patients becoming non-patients, newly trained healthcare staff entering the labour market), the interactions between these networks are changing over time, and each sub-network influences other sub-networks (Aldhous, 2008; Liu et al., 2011).

Health systems research aims to better understand health governance in a context characterized by a multitude of diverse actors (Marsden, 1990). Lebel et al. (2006; WHO, 2009) proposed a conceptual framework to describe the three main characteristics featuring the 'good' governance of socio-ecological systems. These three characteristics can be applied to the governance of health systems: (1) capacity to engage effectively with and handle multiple- and cross-scale dynamics, (2) capacity to anticipate and cope with uncertainties and surprises, and (3) capacity to combine and integrate different forms of knowledge (Table 7.3). These three system properties can be analysed by using five different network properties: two properties related to the structure of the network and three properties related to the position of actors.

Table 7.3 Features identified as important for the adaptive management of resources and the ways in which they are linked to social network structure

Characteristics of health systems	Corresponding social network variables
Capacity to engage effectively with and handle multiple scales	*Reachability*: a measure that describes the capacity to reach many actors to get access to or circulate information (Oh et al., 2004)
	Distance: the shorter the distance between actors, the faster the diffusion of information. Distance between actors is calculated by the number of ties separating two actors
Capacity to anticipate and cope with uncertainties and surprises	*Centrality*: a network with a central structure has more capacity to coordinate actors and provide a rapid response (Oh et al., 2004)
	Betweenness: a rapid response can only be achieved when actors are quickly informed of events or shocks. This requires close links between actors to quickly diffuse information (Fujimoto et al., 2009; Leavitt, 1951)
Capacity to combine and integrate different forms of knowledge	*Reachability*: diversity of knowledge can be achieved through relationships with actors that belong to other spheres or other sub-networks (Granovetter, 1973; Reagans and McEvily, 2003)
	Density: actors in a dense network have more difficulties accessing diverse forms of knowledge, as most actors have very similar backgrounds and values (Steel and Weber, 2001)

How to do social network analysis

A basic approach to SNA captures all three levels of the network during the following three stages: (i) describing the set of actors in the network, (ii) characterizing the relationships between actors, and (iii) analysing the structure of the systems (Table 7.4).

Stage 1: describing the list of actors and members of the network

The first stage of SNA consists of identifying the actors in the network. Some networks have a knowable universe of nodes that the researcher knows in advance (e.g. a list of hospitals in a province). For many other networks, the nodes are not known in advance, as is the case for most policy-making networks or other networks or systems involving largely informal processes. Indeed, a primary goal of many network studies is to identify the members of a network.

In cases where the system has a universe of known actors, researchers can identify those actors by combining two different but complementary methods (Brugha and Varvasovsky, 2000; Grimble and Wellard, 1997): (1) the list of actors involved in a system can be pre-defined by the researcher based on a detailed

Table 7.4 The three main stages of social network analysis applied to health systems research

Stage 1: Defining and/or identifying the network's actors

Roster approach	Free-recall name-generator approach
Step 1: List all actors involved in a system based on a detailed review of documents (roster approach) Step 2: Complement the list of actors with information collected through interviews with key respondents	Step 1: Identify at least one actor involved in a system as a starting point Step 2: Use a name-generator question to generate names of other actors during surveys or interviews. Each name is added to the list and surveyed/interviewed (i.e. respondent-driven or snowball sampling)

Stage 2: Defining the relationships between actors

Roster approach	Free-recall name-generator approach
Step 1: Display the list of actors in a table Step 2: Interview key informants to identify the relationships between actors Step 3: Indicate in the table the existence or absence of a relationship between actors. In each square of the table, a '0' is written when there is no relationship. The square is filled with '1' when there is a relationship between the two actors	Step 1: For each name provided in the survey/interview, follow with a question to identify the relationships between actors Step 2: (Optional) For each relationship identified, ask the interview respondent to indicate the strength of that relationship

Stage 3: Analysing the data (see pages 124–7 for approaches and tools for network data analysis)

review of project proposals and documents; and (ii) this list of actors is complemented by information collected through interviews with key informants.

In cases where the actors in the system are not known, researchers can use the free-recall name-generator approach to identify a list of actors and their relationships. Starting with one or more actors known to be in the system, the researcher surveys or interviews them to ask whom else they share the relationship(s) of interest with. Using a 'snowball sampling' or respondent-driven sampling approach, each of these names is then surveyed with the same name-generator question. The researcher must empirically define the boundary of their network; a common decision rule is to stop eliciting new names when a round of surveys generates fewer new names than the previous round (Lewis, 2006).

The use of roster or free-recall approaches on the same network will generate different results. A roster approach tends to generate much denser networks, as the list of names enables a very low threshold for identifying a relationship. The free-recall approach tends to identify fewer actors who are more diverse and thus tends to create sparser, decentralized networks. A hybrid approach may be preferable in many cases (Henry et al., 2012). However, it is important to pay attention to the data collection approach when interpreting or comparing network findings.

To increase the validity of findings and reduce the incidence of recall biases (e.g. making sure that no actor or part of the system is omitted), a third step can be added: every interviewee is asked to identify additional actors on the basis of their answers to the following questions adapted from Salam and Noguchi (2006): (a) 'who gained or lost during the health intervention?' and (b) 'who is expected to gain or lose as a result of the health intervention's success?' A final set of actors involved in the project was established.

Stage 2: characterizing the relationships between actors

Relationships between actors can be of different kinds and the relationship you choose to measure should be theoretically driven and related to your research question (Salam and Noguchi, 2006). The second stage of SNA consists of identifying the relationships that exist between actors. This information is typically collected at the same time as Stage 1, whether during an in-person interview or through a survey. In the roster approach, the respondent is typically provided with a list of the known actors with opportunities to indicate whether or not they had a given type of relationship with each one (Bodin et al., 2006; Manring, 2007). A similar process is followed for non-roster approaches, the only difference being that the names in the latter approach are generated during the interview or survey.

In an approach used in Ghana, a paper card is created for each actor (i.e. one card per actor). The paper cards are displayed in front of each interviewee and they are asked about the demand for information using the question: 'Do you receive information from this actor?' If the interviewee answers 'yes', the investigator asks additional questions to collect more qualitative information about the type of information received; for example, 'What kind of information do you receive? What is the frequency of your contacts? How do you receive the information – by phone, through visits, letters . . .?'

The same questions are systematically asked about every actor identified. Once completed, the investigator starts again at the beginning of the pile of cards (or the table) and asks about the supply of information: 'Do you provide information to this actor?' If the answer from the interviewee is 'yes', the investigator asks more questions about the type of information provided and the way the information is circulated.

Data collected in Stages 1 and 2 can be recorded in a number of ways. Network researchers often use matrices to organize the existence of relationships, and possibly their strength, between actors in the network (Brinkerhoff, 2004). One matrix records the demand for information and a second one records the supply of information. Each respondent thus generates a row of zeros and ones for each of the two network relations (demand and supply of information): '1' symbolizing the existence of demand/supply of information and '0' signifying no information flow between the two actors. In the example given in Table 7.5, there is circulation of information between all the actors except between the user and the regional directorate and between the user and the regional doctor.

Data can also be entered in an 'edgelist' format, which records the node, the alter, and then additional ties (Table 7.6). Network surveys or interviews can also include questions on node-level attributes, behaviours or outcomes, particularly if they might either explain or be explained by network structure.

Table 7.5 Example of information flow matrix showing the circulation of information between actors

Circulation of information between actors listed in column 1 with actors listed in row 1	User	Regional directorate	Regional doctor	Hospital manager
User		0	0	1
Regional directorate			1	1
Regional doctor				1
Hospital manager				

Table 7.6 Example of information provision edgelist

From	To	Type of information
Hospital manager	User	Treatment guidelines
Regional directorate	Hospital manager	Treatment guidelines
Regional directorate	Regional doctor	Treatment guidelines
Regional doctor	Hospital manager	Patient data

Stage 3: analysing the structure of systems

The third stage in SNA is the analysis itself. The choice of analytic approach depends on the research question and options are constantly growing alongside methodological and computational advances. Using mathematical algorithms (Borgatti et al., 1999) and software (e.g. UCINET, Pajek, R), researchers can calculate node-, tie-, and network-level properties (see Table 7.2 for common network measures).

Cleaning data and missing data

Missing data is more problematic for network data than for traditional data because of the effect that one missing node has on the number of missing ties. Among networks, analysis of directed networks is at greater risk of biased results due to missing data. There are no easy solutions for dealing with missing network data. Instead, researchers should pursue network data collection and analysis only if they are confident they can access and survey/interview most of the network (Knoke and Yang, 2008).

The extent of data cleaning required depends on the data collector's approach and tools. One common purpose of data cleaning in SNA is to ensure that a single person or organization is not represented by multiple entries due to misspellings or different specifications of the same name.

In this example, two matrices are generated: one matrix for demand for information and one matrix for supply of information. The two matrices are then combined to generate a single matrix (Wasserman and Faust, 1994). The final matrix is the result of the addition of the two links. In summary, the new link is N = A + B: A and B being the value of the link in the matrix demand and the matrix supply. The new network is transformed into a symmetrical and dichotomized network (i.e. without direction of links and no strength, just zeros and ones). The new matrix of the system is inserted into the software UCINET, which helps analyse the properties of the network as described in the following section.

The final matrix generated through interviews is analysed with the software UCINET (Freeman, 1979). These calculations are run by the software. The network measure calculated is then analysed to understand how health systems are governed. For example, in order to be able to find multi-scale solutions to multi-scale problems, the actors of a network need to be able to gain access to information from various types of actors, not only from their close colleagues and neighbours. This means accessing stakeholders who have access to different sources of information and different types of power (Borgatti et al., 1999). Access to various sources of information requires a high level of reachability and short distances between actors.

Descriptive analysis

Many researchers start and stop analysis at the descriptive stage. Any of the node, edge or network metrics in Table 7.2 can be described and graphed. Network graphs, or maps, are visually appealing and useful for encouraging discussion (Figure 7.1).

Probabilistic and Bayesian models

The analysis of network data (nodes, ties) cannot rely on traditional statistical estimation frameworks, which assume independence between units of analysis. Instead, units of analysis in network studies – whether nodes or ties – are interdependent of each other and this interdependence must be accounted for. Until recently, it was mathematically and computationally very difficult to do much more than describe the ties between nodes. The introduction of p* models, now commonly referred to as exponential random graph models (ERGMs), has enabled the statistical analysis of ties. EGRMs have been applied to health systems and policy research to explain the exchange of information in health service organizations (Yousefi-Nooraie et al., 2014, 2015) and in health policy networks in Burkina Faso (Shearer et al., 2014). ERGM analyses have generally found that exchanges between actors are better explained by network structure than by the individual attributes of those actors (Goodreau et al., 2009). This has a number of important implications for complexity science. First, the system itself presents an important input, or variable, in the behaviours and outputs of the system. Second, changes in the structure of the system will change how units in the system interact, and thus the output and outcomes of the system.

ERGM is a class of logistic regression model that assumes that two possible network edges are conditionally dependent. ERGM calculates the log odds of a tie between two actors in a network, conditional on the network structure. Independent variables in this equation can be individual-level, as with typical statistical models, or variables that represent network structure, and which are informed by theoretical and empirical evidence demonstrating structural processes in

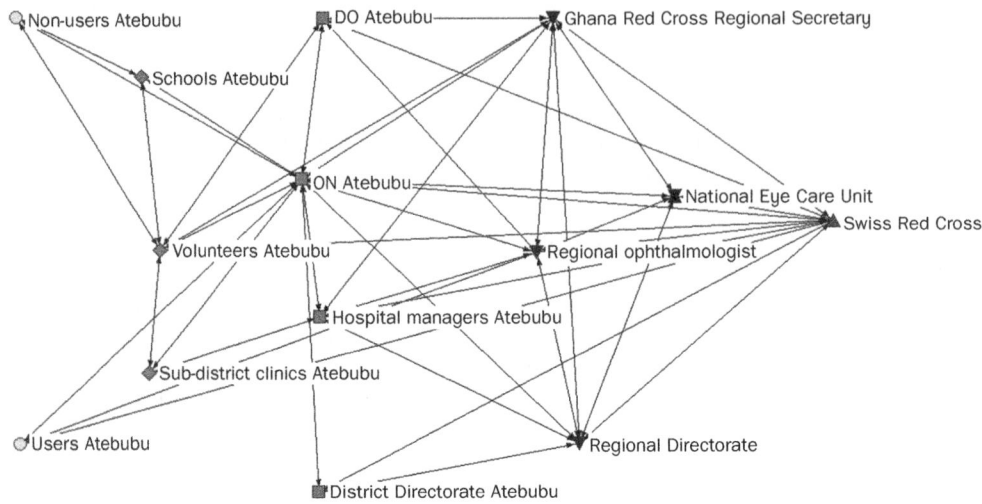

Level of intervention	Name of the actor	Shape of the node
International	Swiss Red Cross	▲
National	National Eye Care Unit	⧗
Regional	Regional Directorate Regional ophthalmologist Ghana Red Cross Secretary	▽
District	District organizers (DO) District Directorate Hospital managers Ophthalmic nurse (ON)	◼
Sub-district	Volunteers Schools Sub-district clinics	◆
Community	Users Non-users	●

Figure 7.1 The sub-network of the district of Atebubu with the six levels of the health system from community to international levels

Source: Karl Blanchet.

networks. These structural processes include transitivity, homophily, and reciprocity (for directed networks).

Models for network dynamics

A cross-sectional snapshot of a network is useful in many ways, but is obviously limited. Both network science and complexity science would agree that a network is not static. Network science can demonstrate theoretically and empirically that these changes are partly a function of structural dependencies, with actor-and tie-level changes in networks depending on earlier actor-, tie-, and network-level characteristics. In longitudinal network models, the dependent variable can be the node or the tie, and models exist to estimate the co-evolution between node-level behaviours and network structure.

Discrete-time network analyses

Researchers who study network change in discrete time steps may be doing so because of the ease of data collection (i.e. a measurement of the network at fixed points in time is often easier to plan and execute) or because they are studying a change in networks that corresponds to fixed events, such as elections, training workshops, etc. Discrete time can be analysed as continuous time though a variety of methods, and indeed continuous-time Markov processes are generally employed for most dynamic network models, regardless of whether networks are measured in discrete or continuous time. The ERGM model described above can also be used to represent discrete-time changes in a network, referred to as a separable temporal exponential random graph model (STERGM). A STERGM is actually formed by two models: one estimating tie formation, the other estimating tie dissolution. Theoretically, it is assumed that a different set of parameters will guide the formation versus dissolution of ties. For example, models that estimate the initiation of collaboration relationships between health services organizations should include a covariate for homophily (by similar geographic location, for example), but one can imagine how this particular attribute might not be as important in a dissolution model. STERGMs are able to deal with multiple observations of the same network over time (i.e. panel data), as well as data that are represented by a cross-sectional network and durational tie information.

Continuous-time network analysis

Stochastic actor-oriented models (SOAM) such as SIENA assume that actors opt to change their network structure, although often influenced by their context, and that ties change one at a time (Snijders et al., 2010). In this way, interpretation of SOAMs occurs at the node level, where the model predicts the probability that actor i will change his or her relationship with actor j (Snijders et al., 2010). In addition, forms of these models can account for observed mutually occurring changes, or co-evolution, in network structure and actor attributes. An excellent application of SOAMs occurred when Yousefi-Nooraie and colleagues (2015) used the approach to determine whether a knowledge translation intervention in a health department changed the structure of the information-seeking networks. In their analysis, the knowledge translation intervention acted to create hierarchies and clusters around individuals who held the new knowledge, instead of decentralizing the information-seeking network (which would be a more efficient structure for information flow). Desmarais and Cranmer (2012) suggest a set of theoretically grounded decision points for determining whether to use a SOAM or STERGM model.

Case study 1

In 1997, Ghana implemented a health sector reform leading to the decentralization of health services, giving more responsibilities to district and regional health authorities and facilities (Sakyi, 2008). In 2010, Ghana had 10 administrative regions. Brong Ahafo, one of these administrative regions, and where this study took place, had 19 districts. The Swiss Red Cross funded an eye care programme with a strong focus on cataract surgery for 10 years between January 1996 and December 2006. The project was jointly implemented by the Ghanaian Ministry of Health and the Ghana Red Cross Society, and was implemented in 12 districts of the Brong Ahafo region. In order to capture the evolution of relationships between actors after the departure of the Swiss Red Cross, an SNA methodology was adopted following all the steps described above.

Method

The study was conducted in five of the 12 districts in Brong Ahafo where the Swiss Red Cross-supported eye care programme was run between 1996 and 2006. Once the hospitals were selected, we focused on the ties between actors that were described in terms of flows of information between all the actors who were involved in the delivery of eye care services. Information was defined in two ways: (1) any piece of information shared between two actors that helped one of the actors or both actors to conduct their day-to-day tasks, and (2) any piece of information that concerned more long-term and strategic issues (Lewis et al., 2008). Face-to-face interviews were conducted in Ghana by the first author. In order to capture the dynamics of social networks, the same people were interviewed at three points in time as suggested by Lewis et al. (2008): December 2006, July 2008, and January 2010. In December 2006, we conducted interviews with key informants to identify the links between the actors of the network. The interviews conducted in July 2008 consisted of verifying the accuracy of the information collected during the year 2006. The actors of the social network were interviewed. The same people were interviewed in January 2010 to find out more about the social network as it was one month earlier in December 2009. A valuable method for generating self-reported ties, which was applied during the interviews, is to use recall lists (Marsden, 1990): a list of all organizations in the field with adjoining empty columns in which respondents could mark their different relations to others (Diani, 2003).

Using mathematical algorithms (Marsden, 1990) and software (e.g. UCINET), researchers have analysed how patterns of relationships between actors within a system can facilitate or constrain both the individual decisions and actions of actors, as well as system functions and adaptive capacities (Wasserman and Faust, 1994).

Results

Between 2006 and 2009, the social network of the eye care system in the Brong Ahafo region of Ghana lost its main broker, the Swiss Red Cross. The structure of the network became more centralized and less dense, which made the circulation of information slower in 2009 than in 2006. The relationship between hospital managers from different districts broke down completely in 2009. The high density of the 2006 network implied that the flow of information was more fluid in 2006 than in 2009. The 2009 network had the shape of a star with five branches, where each branch represented one of the five districts.

Actors in a dense network such as the 2006 network have difficulties accessing diverse forms of knowledge, as most actors have similar backgrounds and values (Frank and Yasumoto, 1998; Granovetter, 1973), which has an impact on the property of good governance defined in the framework. The less centralized structure of the network in 2006 also helped the brokers control information and distribute it across scales (third property of good governance: capacity to combine and integrate different forms of knowledge). The position of brokers at the centre of the 2006 network, combined with their capacity to reach actors from different levels of the network, had a positive influence on the flow of information. From an SNA perspective, the system in 2006 was better governed than in 2009. The changes that occurred between 2006 and 2009 also involved changes in roles and powers between hospital managers and nurses. The 2009 network illustrated the isolation of districts from each other, and the challenges faced by hospital managers in bringing appropriate solutions to potential shocks: specifically, regional and national shocks. However, the 2009 system appeared to district hospital managers as more manageable and easier to control. Their change of position in the network helped them control the circulation of information and get access to information to make decisions and respond to shocks.

Case study 2

Statistical network modelling was used in Burkina Faso to explain why certain policy actors exchanged research evidence for use in policy-making. In studies of factors that support the evidence-informed policy-making, a common determinant is the existence of relationships between research producers and research users. However, despite the identification of this social exchange process, social network analysis has rarely been used to study evidence exchange and use in policy-making. Thus, we mapped the networks of evidence provision and request relationships among policy actors in three health policy decision-making processes at the national level in Burkina Faso and performed network modelling to identify whether actor-level factors (i.e. individual attributes) or network-level factors (i.e. social structure) explained the provision and request of research evidence in health policy-making networks in Burkina Faso.

Method

Policy-making in low- and middle-income countries is characterized by a mix of formal and informal processes and includes a range of actors from different administrative levels and from public, private, civil society, and international sectors. Thus, a roster approach is not well suited to listing actors involved in a given case of policy-making, particularly if the researcher is interested in answering questions about hidden or informal sources of information, influence or power.

This study used a name-generator approach verbally administered during in-depth interviews. The data collector asked each respondent that they interacted with on the policy process in question. For each actor's name provided, the data collector then asked whether they provided or requested research evidence to or from them. Demographic and job-related data were also collected for each actor during the interview. Each actor was contacted for an interview and the same procedure carried out. We used the same decision rule as Lewis (2006) and stopped additional data collection after the fourth round of data collection, at which point fewer new names had been provided in that round than in the previous round. Network data were entered from the paper survey into a matrix in Microsoft Excel.

Analysis

The research question was tie-level, thus suggesting the application of an ERGM model. Network matrices of directed ties were imported into R, and the statnet suite of packages was used to describe the data and build ERGM models, where the dependent variable was the existence of a *provision* or *receipt* of research evidence tie. Following Goodreau et al. (2009), we tested attribute-only, network-only, and combined models to observe the added value of network-level covariates. In the attribute-only model, we included covariates for the actors' organizational affiliation, job level, and whether they had research experience. We also included whether each of these attributes was shared by the two nodes (dyad) in the tie (i.e. homophily). Network, or structural, covariates included transitivity, reciprocity, and layering or multiplexity – the existence of multiple different types of relationships between a dyad. Transitivity, or the tendency to form triangles, indicates closed and cohesive groups of like-minded people. There is a strong social tendency for people to form triangles, but we hypothesized it would actually limit the motivation to seek and use research evidence for evidence-informed policy-making.

Results

Across all policy cases, provision networks were denser than request networks, meaning that evidence was provided more than requested. We observed the greatest number of triangles in the child health network. ERGM models tended to fit best when they combined the attribute and structural effects. In the final models, directed ties were explained by multiplexity but there was no evidence of reciprocity for either provision or request ties (i.e. it was not provided in both directions, or requested in both directions, between a pair of actors). In the child health policy network, the odds of a provision tie were much higher if that tie closed a triangle. This finding was consistent with the qualitative data indicating the closed and cohesive policy paradigm in this network. Overall, the structural findings suggested that evidence exchange is quite hierarchical in Burkina Faso policy-making.

Respondents were also interviewed about their use of evidence to inform policy-making, and reported use correlated well with their number of ties in the network, particularly their provision ties.

Conclusions

The findings of this study suggest that (1) evidence exchange, like many aspects of policy-making, is a social process that depends as much on social structure as individual attributes, and (2) interventions to support the use of evidence in policy-making could draw on network theory and analysis.

References

Aldhous, P. (2008) How the MySpace mindset can boost medical science, *New Scientist*, 2656: 26–7.

Asch, S.E. (1953) Effects of group pressure upon the modification and distortion of judgements, in D. Cartwright and A. Zander (eds.) *Group Dynamics: Research and theory* (pp. 151–62). Evanston, IL: Row, Peterson.

Batley, R. and Larbi, G. (2004) Changing views of the role of the government, in *The Changing Role of Government: The reform of public services in developing countries* (pp. 1–30). Basingstoke: Palgrave Macmillan.

Bodin, O., Crona, B. and Ernstson, H. (2006) Social networks in natural resource management: what is there to learn from a structural perspective?, *Ecology and Society*, 11 (2): 55–62.

Borgatti, S.P. and Cross, R. (2003) A relational view of information seeking and learning in social networks, *Management Science*, 49 (4): 432–45.

Borgatti, S.P. and Foster, P.C. (2003) The network paradigm in organizational research: a review and typology, *Journal of Management*, 29 (6): 991–1013.

Borgatti, S.P., Everett, M.G. and Freeman, L.C. (1999) *UCINET 6.0, Version 1.00*. Natick, MA: Analytic Technologies.

Borgatti, S.P., Mehra, A., Brass, D.J. and Labianca, G. (2009) Network analysis in the social sciences, *Science*, 323 (5916): 892–5.

Brinkerhoff, D. (2004) Accountability and health systems: toward conceptual clarity and policy relevance, *Health Policy and Planning*, 19 (6): 371–9.

Brugha, R. and Varvasovszky. Z. (2000) Stakeholder analysis: a review, *Health Policy and Planning*, 15 (3): 239–46.

Burt, R.S. (2003) The social capital of structural holes, in M.F. Guillen, R. Collins, P. England and M. Meyer (eds.) *The New Economic Sociology: Developments in an emerging field* (pp. 148–89). New York: Russell Sage Foundation.

Christakis, N.A. and Fowler, J.H. (2007) The spread of obesity in a large social network over 32 years, *New England Journal of Medicine*, 357 (4): 370–9.

Cumming, G.S., Bodin, O., Ernston, H. and Elmqvist, T. (2010) Network analysis in conservation biogeography: challenges and opportunities, *Diversity and Distributions*, 10: 414–25.

Degenne, A. and Forsé, M. (1999) *Introducing Social Networks*. London: Sage.

Desmarais, B.A. and Cranmer, S.J. (2012) Micro-level interpretation of exponential random graph models with application to estuary networks, *Policy Studies Journal*, 40 (3): 402–34.

Diani, M. (2003) Networks and social movements: a research programme, in M. Diani and D. McAdam (eds.) *Social Movements and Networks: Relational approaches to collective action* (pp. 299–319). Oxford: Oxford University Press.

Folke, C., Carpenter, S. and Elmquist, T. (2002) Resilience and sustainable development: building adaptive capacity in a world of transformations, *Ambio*, 31 (5): 437–40.

Frank, K.A. and Yasumoto J.Y. (1998) Linking action to social structure within a system: social capital within and between subgroups, *American Journal of Sociology*, 104 (3): 642–86.

Freeman, L.C. (1977) A set of measures of centrality based on betweenness, *Sociometry*, 40 (1): 35–41.

Freeman, L.C. (1979) Centrality in social networks I: conceptual clarification, *Social Networks*, 1 (3): 215–39.

Freeman, M.V. (1984) Role of inside counsel, *Harvard Business Review*, 62 (6): 216.

Fujimoto, K., Valente, T.W. and Pentz, M.A. (2009) Network structural influences on the adoption of evidence-based prevention in communities, *Journal of Community Psychology*, 37 (7): 830–45.

Goodreau, S.M., Kitts, J.A. and Morris, M. (2009) Birds of a feather, or friend of a friend? Using exponential random graph models to investigate adolescent social networks, *Demography*, 46 (1): 103–25.

Granovetter, M. (1973) The strength of weak ties, *American Journal of Sociology*, 78 (6): 1360–80.

Grimble, R. and Wellard, K. (1997) Stakeholder methodologies in natural resource management: a review of principles, contexts, experiences and opportunities, *Agricultural Systems*, 55 (2): 173–93.

Henry, A.D., Lubell, M. and McCoy, M. (2012) Survey-based measurement of public management and policy networks, *Journal of Policy Analysis and Management*, 31 (2): 432–2.

Islam, M. (ed.) (2007) *Health Systems Assessment Approach: A how-to manual*. Arlington, VA: US Agency for International Development in collaboration with Health Systems 20/20, Partners for Health Reformplus, Quality Assurance Project, and Rational Pharmaceutical Management Plus: Management Sciences for Health.

Kiesler, S. and Cummings, J.N. (2002) What do we know about proximity and distance in work groups? A legacy of research, in P. Hiunds and S. Kiesler (eds.) *Distributed Work* (pp. 57–82). Cambridge, MA: MIT Press.

Knoke, D. and Yang, S. (2008) *Social Network Analysis*. London: Sage.

Kohn, L.T., Corrigan, J. and Donaldson, M.S. (2000) *To Err is Human: Building a safer health system*. Washington, DC: Institute of Medicine.

Lazarsfeld, P.F. and Merton, R.K. (1978) Friendship as a social process: a substantive and methodological analysis, in M. Berger, T. Abel and C.H. Page (eds.) *Freedom and Control in Modern Society* (pp. 18–66). New York: Octagon Books.

Leavitt, H. (1951) Some effects of certain communication patterns on group performance, *Journal of Abnormal and Social Psychology*, 46 (1): 38–50.

Lebel, L., Anderies, J.M., Campbell, B. (2006) Governance and the capacity to manage resilience in regional social-ecological systems, *Ecology and Society*, 11 (1): art. 19.

Lewis, J.M. (2006) Being around and knowing the players: networks of influence in health policy, *Social Science and Medicine*, 62 (9): 2125–36.

Lewis, J.M., Baeza, J.I. and Alexander, D. (2008) Partnerships in primary care in Australia: network structure, dynamics and sustainability, *Social Science and Medicine*, 67 (2): 280–91.

Liu, Y.-Y., Slotine, J.-J. and Barabási, A.-L. (2011) Controllability of complex networks, *Nature*, 473 (7346): 167–73.

Manring, S.L. (2007) Creating and maintaining interorganizational learning networks to achieve sustainable ecosystem management, *Organization and Environment*, 20 (3): 325–46.

Marsden, P.V. (1990) Network data and measurement, *Annual Review of Sociology*, 16: 435–63.

McGuire, G.M. (2000) Gender, race, ethnicity, and networks: the factors affecting the status of employees' network members, *Work Occupation*, 27 (4): 500–23.

Milgram, S. (1967) The small world problem, *Psychology Today*, 2: 60–7.

Moreno, J.L. (1934) *Who Shall Survive? Foundations of sociometry, group psychotherapy, and sociodrama*. Beacon, NY: Beacon House.

Newman, L. and Dale, A. (2005) Network structure, diversity, and proactive resilience building: a response to Tompkins and Adger, *Ecology and Society*, 10 (1): r2.

Nowak, M.A. (2006) Five rules for the evolution of cooperation, *Science*, 314 (5805): 1560–3.

Oh, H., Chung, M.H. and Labianca, G. (2004) Group social capital and group effectiveness: the role of informal socializing ties, *Academy of Management Journal*, 47 (6): 860–75.

Olsson, P., Folke, C. and Hahn, T. (2004) Social-ecological transformation for ecosystem management: the development of adaptive co-management of a wetland landscape in southern Sweden, *Ecology and Society*, 9 (4): 2.

Padgett, J.F. and Ansell, C. (1993) Robust action and the rise of the Medici, 1400–1434, *American Journal of Sociology*, 98 (6): 1259–1319.

Reagans, R. and McEvily, B. (2003) Network structure and knowledge transfer: the effects of cohesion and range, *Administrative Science Quarterly*, 48 (2): 240–67.

Riggan, M. and Supovitz, J.A. (2008) Interpreting, supporting, and resisting change: the geography of leadership in reform settings, in J.A. Supovitz and E.H. Weinbaum (eds.) *The Implementation Gap: Understanding reform in high schools* (pp. 103–25). New York: Teachers College Press.

Sakyi, E.K. (2008) Implementing decentralised management in Ghana: the experience of the Sekyere West District health administration, *Leadership in Health Services*, 21 (4): 307–19.

Salam, M.A. and Noguchi, T. (2006) Evaluating capacity development for participatory forest management in Bangladesh's Sal forests based on d4RsT stakeholder analysis, *Forest Policy and Economics*, 8: 785–96.

Shearer, J.C. (2015) Policy entrepreneurs and structural influence in integrated community case management policymaking in Burkina Faso, *Health Policy and Planning*, 30 (suppl. 2): ii46–ii53.

Shearer, J.C., Dion, M. and Lavis, J.N. (2014) Exchanging and using research evidence in health policy networks: a statistical network analysis, *Implementation Science*, 9 (1): 126 [doi: 10.1186/s13012-014-0126-8].

Snijders, T.A.B., van de Bunt, G.G. and Steglich, C.E.G. (2010) Introduction to stochastic actor-based models for network dynamics, *Social Networks*, 32 (1): 44–60.

Spillane, J.P. (2006) *Distributed Leadership*. San Francisco, CA: Jossey-Bass.

Steel, B.S. and Weber, E. (2001) Ecosystem management, decentralization, and public opinion, *Global Environmental Change A: Human and Policy Dimensions*, 11 (2): 119–31.

Thompson, G.N., Estabrooks, C.A. and Degner, L.F. (2006) Clarifying the concepts in knowledge transfer: a literature review, *Journal of Advanced Nursing*, 53 (6): 691–701.

Uzzi, B. and Gillespie, J.J. (2002) Knowledge spillover in corporate financing networks: embeddedness and the firm's debt performance, *Strategic Management Journal*, 23 (7): 595–618.

Valente, T.W. and Pumpuang, P. (2007) Identifying opinion leaders to promote behavior change, *Health Education and Behavior*, 34 (6): 881–96.

Wasserman, S. and Faust, K. (1994) *Social Network Analysis: Methods and applications*. Cambridge: Cambridge University Press.

Watts, D.J. and Strogatz, S.H. (1998) Collective dynamics of 'small world' networks, *Nature*, 393 (6684): 440–2.

Webb, C. and Bodin, O. (2008) A network perspective on modularity and control of flow in robust systems, in J. Norberg and G.S. Cumming (eds.) *Complexity Theory for a Sustainable Future* (pp. 85–118). New York: Columbia University Press.

World Health Organization (WHO) (2009) *Scaling Up Research and Learning for Health Systems: Now is the time*. Geneva: WHO.

Word Health Organization (WHO) (2010) *Strengthening Health Systems: What works?* Alliance for Health Policy and Systems Research Annual Report 2009. Geneva: Alliance for Health Policy and Systems Research/WHO.

Yousefi-Nooraie, R., Dobbins, M. and Marin, A. (2014) Social and organizational factors affecting implementation of evidence-informed practice in a public health department in Ontario: a network modelling approach, *Implementation Science*, 9: 29 [doi: 10.1186/1748-5908-9-29].

Yousefi-Nooraie, R., Dobbins, M., Marin, A., Hanneman, R. and Lohfeld-Uzunoz, E. (2015) The evolution of social networks through the implementation of evidence-informed decision-making interventions: a longitudinal analysis of three public health units in Canada, *Implementation Science*, 10 (1): 166 [doi: 10.1186/s13012-015-0355-5].

8 | Human systems dynamics: a tool for understanding self-organizing behaviour of actors in the system

Glenda H. Eoyang

Introduction

Health, health care, and healthcare systems involve complex relationships and processes, many of which are difficult, if not impossible, to predict or control. Fluid boundaries, multiple co-dependent variables, and massive interdependencies challenge traditional methods of data collection, analysis, action, and evaluation. These conditions result in radical uncertainty of pathways and outcomes in self-organizing systems. They generate challenges that have no rational or stable solution, but that does not mean there are no useful ways forward. These problems qualify as 'wicked problems', and they require a special set of methods and tools (Rittel and Webber, 1973) that support iterative, adaptive theory and practice (Eoyang, 2011). The sciences of complexity and non-linear dynamics offer ways to understand and describe the mechanics of such complex, emergent systems (Boulton et al., 2015). Human systems dynamics draws from that theory base to inform decision-making and action-taking in human systems at all scales, including individual, team, organization, and community.

The remaining sections of this chapter will explain what human systems dynamics is and how it can be important for health systems research as well as introducing the approach, the foundational methods, and some of the tools that support the work. Practical considerations and tips for good practice will follow, and the chapter will end with a case study of health systems dynamics applied to a variety of challenges in healthcare settings. The purpose of the chapter is to prepare individuals to apply the principles and practices immediately to their own complex health system challenges.

Human systems dynamics principles

Human systems dynamics (HSD) is based on the assumption that human systems self-organize. Any change within a system, whether at the individual, team, organization or community level, is motivated by internal tensions. These tensions form the dynamics of the system that influence the potential for and activation of change processes at every scale of interaction. While these emergent processes of self-organization are unpredictable and uncontrollable, they can be observed, anticipated, and influenced. A fundamental understanding of theory of non-linear dynamics and practice in emergent environments has led to a suite of models and methods that help researchers and practitioners to see, understand, and influence performance in complex systems. In this chapter, we will explore six of the tools that are most relevant to research in medical systems: Container-Difference-Exchange (CDE) model for self-organizing in human systems, Pattern Logic, Adaptive Action, Same & Different, landscape diagrams, and designing exchanges. We will begin with the CDE model, which sets the foundation for all of the others.

Based on our research, three conditions establish the speed, path, and outcomes of self-organizing processes: containers, differences, and exchanges (Eoyang, 2001).

Containers establish the boundaries within which self-organizing patterns emerge. In human systems, the containers are multiple and massively entangled. In healthcare systems, they can include individual physical, emotional, and mental states; families; teams of care providers; facilities; communities; or any other physical, social or cognitive boundary. Any one of these creates a unit of analysis or action (a container) that influences behaviour in the system as well as changing in response to systemic behaviours over time.

Differences within the container carry information about the current state of the system. They also create the potential for change as they accumulate, hold, and release tensions in the system. Traditional dependent and independent variables are examples of differences, but many others can also influence the self-organizing processes in the system: power; general wellbeing; levels of engagement; bias, trust; income, dress, race and ethnicity, and so on are other differences that can make a difference as a human system self-organizes.

Exchanges connect parts of the system to each other and allow the tension of difference to increase or be released. Flows of resources, information, policy, feedback, physical connection, and delivery of care are all exchanges that resolve differences and so influence emergent change in healthcare settings.

Together, these three conditions – container, difference, exchange – constitute the Eoyang CDE model for self-organizing in human systems. A system's containers, differences, and exchanges are interdependent, and they influence the speed, path, and outcomes of self-organizing systems. When these conditions are highly constrained (small containers, few differences, and tight exchanges), the self-organizing process can be relatively predictable and controllable. We observe high constraint of the conditions, for example, when a sample size is small (constrained container), variables are few and unambiguous (constrained differences), and relationships are reliable and strong (constrained exchanges). Under conditions such as these, randomized control trials and other traditional methods are appropriate because the conditions are constrained and systemic behaviours can be predicted and controlled.

In contrast, when these conditions are unconstrained (large containers, many differences, and loose exchanges), self-organizing can be unpredictable and impossible to control. Such conditions exist, for example, in communities (unconstrained containers), when variables are many and co-determinant (unconstrained differences), and power relationships are weak (unconstrained exchanges). In these situations, traditional research methods are not appropriate because the systems are unpredictable and uncontrollable, so standards of replicability, reliability, and validity do not apply.

The practice of HSD provides methods to see, understand, and influence these fundamental conditions of self-organizing processes for individuals, teams, organizations, and communities (Eoyang and Holladay, 2013). The CDE model informs design and execution of research in three ways.

First, in a particular system, data can be collected regarding the boundary conditions (C), the relevant variables (D), and the connections between and among systemic entities (E). These data may include expected and obvious measures, as well as items that are qualitative and/or unexpected. For example, meeting designs can provide information about a team. Containers would be the size and arrangement of the location, the amount of time, and the number of members. Differences would be the number and novelty of items on the agenda. Exchanges would be the frequency and diversity of interactions before, during, and after the meeting.

Second, the CDE model provides insights to support analysis of data. Change in one of the conditions influences the others because the three conditions are interdependent. The relative constraints of the conditions can provide information about the potential for change in the future. For example, if the frequency and diversity of interactions diminishes as a group meets over time, then analysis questions would focus on changes in the containers and/or differences that might have caused the shift.

Finally, the CDE model informs action because intentional interventions to shift the containers, differences or exchanges will influence all conditions, and shift the patterns of and relationships among the conditions that were observed as well as others that might be interdependent but not deemed relevant in the original data collection. In the meeting example, additional stakeholders might be invited to increase differences in the hopes of increasing the quality and quantity of interaction. This intervention would also increase the container and possibly introduce new systemic features of interest.

Given this understanding of conditions for self-organizing in relation to complex dynamics, the theory and practice of HSD are based on two fundamental models and methods: Pattern Logic and Adaptive Action.

Pattern Logic

Pattern Logic represents a system in terms of its self-organizing capacity. The pattern logic model supports understanding and action in three ways. First, the model describes the conditions for self-organizing (CDE) that influence the state of the system for a particular place, time, and purpose. Second, it supports assessment of future CDE possibilities by recognizing the tensions that are held in the current patterns. Finally, the model informs action to influence the conditions and, therefore, the future path of the self-organizing process.

Many different methods can be used to identify and analyse the containers, differences, and exchanges of the pattern logic for self-organizing. The choice of an appropriate method depends on the data available, the dynamics of interest, and the skills and discourses of stakeholders. For example, we have found various systems mapping approaches useful to model CDE. In our experience, a systems dynamics model captures containers as stocks, differences as parameters, and exchanges as flows (Meadows and Wright, 2008). An agent-based model captures containers as agents, differences as characteristics of those agents, and exchanges as rule-based interactions (Miller and Page, 2007). A network model of a system represents containers as nodes and hubs, differences as features, and exchanges as edges (Barabási and Albert, 1999). Any of these approaches can reveal the pattern logic of the system (Eoyang, 2015). Two modelling methods are particularly designed for use in HSD: CDE portraits and bubble diagrams (Eoyang, 2014).

CDE portraits and bubble diagrams

Any complex system will include multiple containers at many different scales, and each container has within it relevant differences and exchanges. It would be impossible and unnecessary to capture them all in a systems model. A useful model will include features of the system that are relevant for a particular purpose. A *CDE portrait* is a tool to capture and document those relevant features.

Data regarding the CDE conditions can come from individual reflection and reporting, interviews, small group discussion, observation or document review and analysis, depending on the question to be addressed and project constraints. The process is relatively simple and the model is necessarily incomplete. The process can be repeated to generate multiple CDE portraits until one generates meaningful understanding and options for action.

1. The pattern of interest is identified.
2. Three containers related to that pattern are selected.
3. Within each of the three containers, three relevant differences and three meaningful exchanges are identified.

An example from a recent unpublished healthcare intervention is shown in Table 8.1. The challenge was that some parents complained that they were not included in discussions and decisions about the care of their children in the emergency room at the hospital. The box provides a simplified model of the relevant aspects of the challenge to be addressed.

The same content can be presented graphically as in the *bubble diagram* in Figure 8.1. Both of these modelling methods are particularly useful because they are simple to understand and generate. They capture the pattern logic of current conditions that are relevant to the challenge at hand. These models provide information about the system to inform adaptive action, which is the second principle of HSD.

Adaptive Action

A complex adaptive system is constituted by semi-autonomous agents that are free to act in unpredictable ways and whose interactions generate system-wide patterns (Dooley, 1996). Change in complex adaptive systems is perpetual and unpredictable. Patterns, determined by the conditions for self-organizing, shift as tensions accumulate and release among containers, differences, and exchanges. In addition, conditions interact within and between patterns that are massively

Table 8.1 CDE portrait

Containers	Differences within	Exchanges within
Emergency room	Expertise of staff	Triage
	Times of day	Intake
	Age of patient	Deliver treatment
Physician team	Specialty	Case review
	Familiarity to staff	Scheduling
	Interest in collaboration	Social engagement
Parent	Age of patient	Connect with doctor
	Parents' attitudes at release	Provide information
	Parents' attitudes at intake	Get follow-up instructions

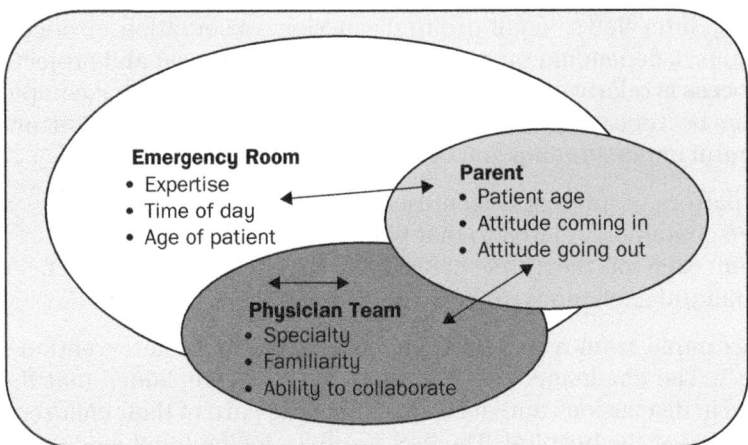

Figure 8.1 Pattern analysis: emergency room case example

entangled. For example, in the previous scenario, interactions among physicians might shift their relationships with parents, as parents' interactions with each other might influence and be influenced by relationships with emergency room personnel.

Because of these open, high-dimension, and non-linear self-organizing change processes, traditional methods of data collection and analysis are not always feasible for a variety of reasons. Open boundaries can make it impossible to determine a stable, representative sample population. The number and co-dependency of variables may make traditional data collection and analysis processes meaningless. Non-linear causal relationships can introduce levels of co-variance that require non-linear time-series analysis or other unusual statistical approaches. Under these circumstances, non-traditional methods are required to make meaning out of systems that are unpredictable and uncontrollable. In particular, methods to study self-organizing processes must be iterative, deal with unstable relationships, and allow for an unlimited and/or indeterminate number of variables. In HSD, Adaptive Action is the iterative inquiry cycle that meets these requirements because it supports intentional engagement with complex systems as they self-organize. Adaptive Action consists of three questions:

1. *What?* This question captures information about the current pattern of self-organizing conditions of the system. While any data collection method can be useful here, Pattern Logic of the CDE model can be particularly helpful. What are the most relevant containers? What differences make a difference? What exchanges support change or hold the system in a stable state?

   Any method of data collection can fit into this step of Adaptive Action. Any source of information can provide evidence of an emerging pattern: surveys, focus groups, observation, document review, measurement, stories. Whatever gives you intelligence about the patterns emerging from the chaos can help you determine the What? for your Adaptive Action.

2. *So what?* This question begins the analysis of the current pattern. Helpful questions might include: So what tensions are present currently? So what is working, or not working? So what are multiple perspectives, options for action, challenges and risks? So what has changed and what might change in future? So what options for action exist to shift the conditions and what are the risks and benefits of each?

   Both traditional and non-traditional analysis methods can be useful in this step of Adaptive Action. Standard statistics are useful if you would expect a normal distribution within the pattern. If not, non-linear analysis, qualitative methods, and thematic approaches can surface the most relevant and significant features to inform options for action.

3. *Now what?* This question is the invitation to move into action. Questions might include: Now what will be done? Now what will be communicated? Now what will be expected as a consequence of the action? The action, whatever it is, influences one or more of the conditions in the system – container, difference or exchange. That shift changes the situation and generates a new pattern, which leads to the next Adaptive Action cycle: *What?* (Eoyang and Holladay, 2013).

   This step of Adaptive Action can draw on any capacity of the system – at any scale. Individual contributors, lone professionals, teams, executives or elected officials can use whatever is in their power to take action to shift the conditions that form the patterns. Traditional planning, accountability, communication, and collaboration, all are important components of effective action in the Now what? phase of Adaptive Action.

The process of Adaptive Action is similar to action research and other inquiry cycles. It can be localized and quick, as in a conversation or a clinical intervention, or more generalized and long term, as in strategic planning processes or longitudinal studies. The key features of Adaptive Action are that the process is iterative, to capture the emergent change in the system, and inquiry-based, to adapt to the uncertainty of the self-organizing processes.

Together, these two fundamental features – Pattern Logic and Adaptive Action – inform every human systems dynamics inquiry and intervention.

Human systems dynamics tools

HSD incorporates a wide variety of tools from psychology, organization development, sociology, education, management, and other fields of human systems research and practice (Poole et al., 2000). Any tools that reveal patterns, assess situations or inform action can be incorporated into Adaptive Action cycles to see, understand, and influence patterns as they self-organize in human systems. Some tools are particularly well-suited or designed for iterative inquiry and pattern-based processes of HSD. The most effective tools support all three steps of the Adaptive Action process by revealing patterns, assessing tensions, and suggesting action to increase or release tensions to influence change. We will explore three commonly used HSD tools here: Same & Different, landscape diagrams, and designing exchanges.

Same & Different

This tool is a simple, flexible option for finding and articulating patterns in complex systems. It may be used by any number of stakeholders, and a facilitated conversation can be designed to meet a wide range of questions and contexts. Data to inform this process can come from any data source, including quantitative measures and trends over time, dialogue with stakeholders, individual observation, and narrative analysis. A typical Adaptive Action design for Same & Different is described next.

Same & Different: What?

A group identifies the context or question to be addressed. Two parts of a system or situation are named. The group brainstorms the ways that the two are the same and ways that they are different. The brainstorm is captured on a simple, two-column table. The columns are titled 'SAME' and 'DIFFERENT', as in Table 8.2.

Table 8.2 Same & Different: What?

Context or question to be addressed: Physician team and parents of teenaged patients

SAME	DIFFERENT
Approximate age	Emotional involvement
Community membership	Professional status
Language spoken	Familiarity with ER procedures
Busy and stressed	Amount of time available for discussion
Concerned about the welfare of patient	Relationship with patient
And so on . . .	And so on . . .

The list of similarities and differences in Table 8.2 constitutes a simplified, Pattern Logic representation of the system. There are no 'right' or 'wrong' answers at this point of the analysis. The lists will not be complete or definitive. Given the emergent complexity of human systems, any two groups might come up with different lists, and the same group might name different things at different times. The key is that everything on the list is relevant to someone, and each voice has an opportunity to add to the list.

Same & Different: So what?

When the brainstorming process is complete, the group reviews the lists and assesses each of the similarities and differences in terms of whether it has a positive (+), negative (−), neutral (0) or unknown (?) influence on the current pattern. This conversation should be facilitated and will uncover different values, assumptions, and perspectives of the group. Items of disagreement do not need to be resolved – they can simply be marked as unknown. Again, there is no 'right' or 'wrong' answer, but the responses should reflect the insights of the group. If a difference of opinion arises, the item is marked with a question mark, and the conversation continues. Table 8.3 provides such an analysis of the example data.

Same & Different: Now what?

The group selects one similarity or difference on which to take action. The focus may be a similarity that should be strengthened or disrupted. For example, the doctor might articulate his or her concern about the patient to make this similarity more explicit. Or, the intervention might magnify or minimize a difference. For example, the doctor might behave differently to minimize the tensions related to status in the ER. Only one item is chosen as a focus of action because even a small change will shift the pattern in unpredictable ways and move the group into the next Adaptive Action. Together, the group decides on a collaborative action that they believe will have the desired influence on the focus feature of the system. The action is taken, and the group reconvenes at some later time to assess the effects and collect data for the next *What?*

Table 8.3 Same & Different: So what?

Context or question to be addressed: Physician team and parents of teenaged patients

SAME	DIFFERENT
0 Approximate age	? Emotional involvement
+ Community membership	− Professional status
+ Language spoken	+ Professional skills
− Busy and stressed	− Familiarity with ER procedures
+ Concerned about patient's welfare	− Time available for discussion
And so on . . .	+ Relationship with patient
	And so on . . .

Landscape diagrams

The landscape diagram is derived from the work of Brenda Zimmerman (Zimmerman et al., 1998). It helps individuals and groups recognize and influence the levels of constraint and stability that are inherent in a human system at a particular time and place and context.

Landscape diagrams can be used to analyse many different kinds of qualitative and quantitative data. Stories, interview notes, proxy measures for agreement and/or certainty can all be excellent sources of data. Usually a group uses the landscape to reflect on and analyse data that represent their situation and status of their decision-making.

A landscape diagram is a two-dimensional plot of agreement and certainty. The origin represents a system close to agreement and close to certainty, and both features decrease as they move away from the origin. Three zones are identified in the landscape diagram. When the system is close to agreement and close to certainty, the pattern is considered to be stable. Far from agreement and certainty, the pattern is unstable or chaotic. In between the extremes, the pattern is actively in self-organizing mode and new patterns will emerge over time.

Agreement and certainty on the landscape diagram represent two patterns that are relevant to group decision-making. When the pattern is tightly constrained (close to agreement and close to certainty), the behaviour of the system is relatively predictable. Best practices and treatment protocols are appropriate in this part of the landscape. When the pattern is unconstrained (far from agreement and far from certainty), the behaviour of the system will be random and chaotic. Emergency response and individual response to chronic disease are characteristic of this zone of the landscape. The conditions of the system can be far from agreement and close to certainty, for example, in situations of interdisciplinary practice where disagreement is strong and perfectly predictable. On the other hand, when the pattern is close to agreement and far from certainty, a team or community may share a deep anxiety about future opportunities and risks.

Landscape diagram: What?

The landscape diagram supports pattern recognition in the *What?* stage of Adaptive Action (Figure 8.2). Individuals or groups describe a particular situation or aspect of a situation in terms of the amount of agreement and certainty they perceive. Using the landscape as a model, a group surfaces and negotiates a shared understanding of the current dynamics in a system. This analysis may be facilitated in any number of ways. For example, individuals can place markers where they think an issue fits on the landscape. Members of a group may write stickers to represent different aspects of a problem and place them where they belong on the landscape. The landscape may be marked on the floor, and individuals and groups can move to the places on the landscape where they believe a particular issue resides currently. Any of these approaches helps a group explore and come to agreement on the current level of stability or instability in the system of interest.

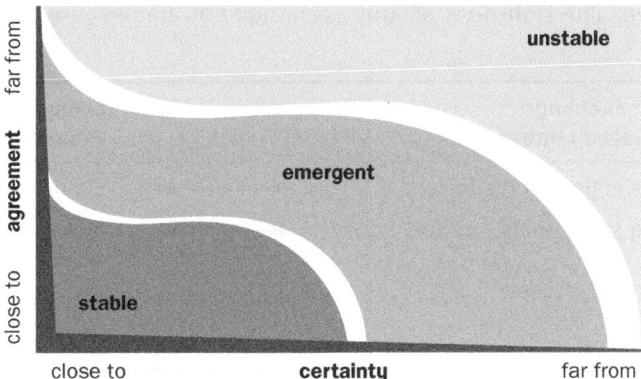

Figure 8.2 Landscape diagram: What?

Source: © Human Systems Dynamics Institute. Reproduced with permission.

Landscape diagram: So what?

After the current state is described and located on the landscape, the group considers whether the current pattern is in the optimal zone for the current situation. They deal with the following questions, as appropriate: Is the system too stable and need to be disrupted to meet current needs? Is it too emergent and need to be constrained into the stable zone or disrupted into the unstable zone? Is the situation too chaotic and unstable, so that it needs to be bounded and settled into a more predictable and reasonable state? Again, this discussion should surface systemic assumptions and perspectives to move a group towards shared mental models and common understanding.

Landscape diagram: Now what?

When the group has completed their analysis and chosen the direction in which they want to influence the system, they consider options for action to shift the patterns in the desired direction. Using Pattern Logic, they consider increasing constraint (i.e. smaller containers, fewer differences, and tighter exchanges) to move the system towards greater stability. Decreased constraints (i.e. larger containers, more differences, and looser exchanges) encourage the system towards greater instability.

Again, the group defines common action, takes it, and returns to analyse outcomes and collect data for the next Adaptive Action cycle.

Designing exchanges

Exchanges include the pattern features that support any flow that connects parts of the system together. The exchanges can pass anything, including information, resources, physical contact, non-verbal communications, or bombs and bullets. In complex systems, the exchanges are critical because they either hold a system in place or activate its movement. Tight exchanges tend to restrict and direct change in a system, and loose ones allow change that is unpredictable and uncontrolled. The tightness of any exchange can be assessed according to four criteria:

	Tight exchange Increased control	◀────▶	Loose exchange Decreased control
Length	Short in time or distance	◀────▶	Long in time or distance
Width	Many simultaneous connections	◀────▶	One or few simultaneous connections
Direction	One-way	◀────▶	Two-way
Dynamic	Amplifying	◀────▶	Damping

Designing exchanges: What?

This step of the Adaptive Action explores what the current situation is with exchanges across the system. A group brainstorms a list of the current exchanges and chooses one that is most relevant to the pattern of interest. They evaluate the relative strength of the exchange by considering each of the criteria in sequence. This process provides a current-state profile of the systemic exchange.

Designing exchanges: So what?

The group then decides whether the exchange needs to be tighter to afford more control or looser to afford more freedom for the system to be fit for its future function. They consider what are the risks and benefits of increasing or decreasing exchange-based control of the system's parts, and they explore options for action to shift the exchange to be more fit for function.

Designing exchanges: Now what?

Finally, the group selects one action that they will implement immediately. The consequences of the action will be assessed and adjusted in their next Adaptive Action cycle. A table is used to collect and analyse the information in each step of this process (see Table 8.4).

Practising human systems dynamics

The principles and tools of HSD are deceptively simple, but doing the work can be a challenge. The fundamental assumptions HSD makes about change are radically different from previous approaches. Attitudes and habits that were successful before can lead to failure in the unpredictable emerging world of self-organizing human systems. Effective practice in human systems dynamics requires practices of humility and inquiry. HSD models and methods have been successfully applied in a wide range of settings, primarily in North America. Applications in the developing world include training and practice in health professions education, youth development, and response to corruption, but those cases are not available for reference at this time. We encourage people to adapt the concepts

Table 8.4 Exchange analysis

Exchange to be investigated: Parents in the Emergency Room (ER)

	What is the current state?	So what are the risks and benefits of this state?	So what might be an adjustment?	Now what will we do?
What is the length of time from message to response? (length)	Short	Parents are under pressure to respond quickly → increased anxiety	Provide an Emergency Room (ER) guide to engage with parents	
What is the information content of the message? (width)	Narrow	No context to help parents support → increased anxiety	Connect in conversations with all professionals, not just the physician	Provide a Parents' Guide to clarify questions and expectations
What is the feedback loop? (direction)	One-way	No questions or discussion → increased anxiety	Allow for questions and dialogue	
Does the exchange encourage or discourage? (dynamic)	Damping	No sign of positive movement → increased anxiety	Identify positive, as well as negative signs for the parent	

and tools to fit the culture and language of researchers and subjects, so multiple versions have been developed for specific settings.

In various settings and applications, the following simple rules have been developed to support the work of Adaptive Action and Pattern Logic:

1. *Focus on fit for function.* No single yardstick measures success in a complex adaptive system. Each situation requires a different set of self-organizing conditions. What is too small a container in one situation may be too large in another. One set of differences may be perfect focus in one group and a major distraction in another. At one time or for one purpose, a tight exchange may be optimal, while at a different time or from a different perspective less constraint would be more useful. The measure of success must always be how well the system is serving its function, and as that function emerges over time, the understanding and action in the system must evolve, too.
2. *Stand in inquiry.* Every complex system emerges continually and in surprising ways. Any assumption, any answer, no matter how wise it may be now, will be totally mistaken in the near future. An effective practitioner holds answers lightly and asks questions often.
3. *Expect to be surprised.* Control and predictability are rare in self-organizing systems. Even when the conditions are highly constrained, and the system is stable in the short term, the situation may shift at any time. A wise practitioner is always aware of the power of the unexpected.
4. *Explore multiple perspectives.* Because the patterns in human systems involve so many different containers, differences, and exchanges, no single interpretation or perspective is likely to tell the complete truth. When one suspends disbelief and sees patterns through a variety of lenses, new options for action emerge spontaneously.
5. *Value variety.* The dynamics of complex systems generate unique patterns at every moment and in every place. Patterns in human systems are always context-dependent, and the differences among situations can reveal hidden value.
6. *Attend to the part, the whole, and the greater whole.* Individuals and groups, teams and organizations, neighbourhoods and communities – every scale of the human system determines and is determined by those that are within and beyond it. HSD inquiry must always focus on multiple scales to understand change in any scale.

HSD tools in action: a case study for strategic Adaptive Action learning teams

We have used Adaptive Action for research and change in human systems at many scales. Individuals and teams receive coaching to adapt to changing circumstances. Organizations use Adaptive Action for transformative change and strategic planning. Communities build collaborative capacity as they engage as individuals and institutions in shared Adaptive Action. The following case study focuses on a cohort of medical professionals who were engaged in a series of Adaptive Action Laboratories to respond to wicked challenges for their teams, patients, and organization.

Ten different teams were involved in the programme. Each one brought a specific, strategic challenge to the programme. Each team worked over an eight-month period to complete multiple Adaptive Action cycles. All of them made substantive breakthroughs in seeing, understanding, and influencing patterns of wicked problems across the system.

What?

The process began with an initial problem statement in the *What?* phase of the Adaptive Action. Each team was invited to describe their challenge in three sentences. Although the issues were systemic and strategic, the three-sentence descriptions captured the essential Pattern Logic to for the first stage of the Adaptive Action.

Dyads were formed across teams, so that each person was engaged with someone from a different team. One member of the dyad shared their three-sentence description, and the other asked questions about the patterns they heard. The speaker was not allowed to respond, but to listen to the questions and consider how each one might shift their perception of their wicked problem. After five minutes and many questions, the members switched role, and the second person shared their story and received questions. At the close, each person reflected and wrote in their journals about what they had heard. The teams reconvened, shared their insights, and revised their three-sentence descriptions. The groups were encouraged to model their initial descriptions as network maps, CDE portraits, or bubble diagrams. These diagrams and the three-sentence stories constituted the *What?* stage of the Adaptive Action cycles.

So what?

This stage of the process was different for each group, as they chose one or another HSD tool to help them make meaning of the pattern they had articulated in the *What?* phase of the process.

Group A focused on a challenge in inter-professional practice among nurses and allied health professionals. They used the Same & Different process to articulate the tensions in the patterns between the two professional groups. After discussing the variety of challenges and tensions, they decided to focus on differences of technical jargon because they recognized problems that arose when each professional group had its own language. The group explored a variety of ways in which they might create new containers and/or exchanges to mediate or moderate the tension caused by professional jargons.

Group B was concerned with developing an evaluation design for a complex programme. They used the landscape diagram to identify aspects of the programme that were more or less stable based on agreement among the stakeholders and certainty of outcomes over time. When aspects of the system had been categorized, the group explored various options for designs to match the relative stability of the programme feature. They identified potential evaluation questions, indicators, data collection and analysis techniques that would be most appropriate for each of the levels of stability.

Group C focused on issues of patient engagement and effective communication. They analysed exchanges between parents of children and emergency room personnel to determine the current levels of constraint and options for increasing the strength of some exchanges and decreasing constraint of others.

The other seven groups chose other HSD models and methods to analyse the tensions in the patterns related to their wicked problems. Each group identified a variety of possible actions and potential costs, benefits, and risks of each.

Now what?

Finally, each of the ten groups assessed their options for action and selected the one action that would be easiest, fastest, and most likely to bring about a positive shift in the pattern.

Group A decided to post a bulletin board where anyone could post terms they did not understand. Others would post definitions and examples in response. After a week, the group reconvened to clarify outstanding issues, evaluate their progress, and begin their next Adaptive Action.

Group B decided to begin their evaluation design with facets of the programme that were the most stable – high agreement and high certainty. Traditional evaluation strategies are most successful in stable systems, so this design was a 'low hanging fruit' for the group. They defined these aspects of the evaluation plan and submitted it for review. At their next meeting, the group began to explore more complex and innovative evaluation strategies to capture and assess patterns that were less predictable in the emergent and unstable zones.

Group C determined that the initial exchanges between parents and ER personnel were too loose to set the stage for reasonable communication later in the process. Professionals were preoccupied with triage and rapid response at the very time that the parents needed clarity and information. The group decided to appoint one person in the response team to meet with the parent immediately upon admittance and to return to give additional information and answer questions every 15 minutes. The question for the next Adaptive Action was to see if this investment was sufficient to strengthen the exchange and create a more stable pattern between the family and the caregivers.

None of these groups could be certain of the outcomes of their Adaptive Actions, but they had plans and clear arrangements to take action, collect information about results, and prepare for adaptations in the next Adaptive Action cycle. Any of the groups might have chosen any number of alternative approaches, but the uncertainty of the system means that a perfect solution would be impossible to find. The best alternative was to use Pattern Logic and Adaptive Action to see, understand, and take action to influence change as it emerged in the complex dynamics of the human system.

Conclusion

Health, health care, and the healthcare system are all complex, self-organizing, and emergent. Using lessons from complexity science, human systems dynamics (HSD) provides principles, tools, and practices to engage productively with human systems that cannot be predicted or controlled.

This chapter has introduced the foundations of HSD and provided examples of how the disciplines of Pattern Logic and Adaptive Action empower responsible decision-making and action-taking. The emergent nature of human systems dynamics will continue to surprise and disrupt even the most robust plans for engagement, but the humility and inquiry of HSD prepare researchers and practitioners to make choices and take action that respond to the needs of the system and shift patterns towards greater fit for function.

References

Barabási, A.L. and Albert, R. (1999) Emergence of scaling in random networks, *Science*, 286 (5439): 509–12.

Boulton, J., Allen, P. and Bowman, C. (2015) *Embracing Complexity*. New York: Oxford University Press.

Dooley, K. (1996) A nominal definition of complex adaptive systems, *The Chaos Network*, 8 (1): 2–3.

Eoyang, G. (2001) *Conditions for self-organizing in human systems*. Unpublished doctoral dissertation, Union Institute and University, Cincinnati, OH.

Eoyang, G. (2011) Complexity and the dynamics of organizational change, in P. Allen, S. Maguire and B. McKelvey (eds.) *Sage Handbook of Complexity and Management* (pp. 319–34). Thousand Oaks, CA: Sage.

Eoyang, G.H. (2014) *Magic 21*, 17 April [retrieved April 14, 2016, from http://wiki.hsdinstitute.org/hsd_magic_21; accessed 14 April 2016].

Eoyang, G.H. (2015) *Escape the Pressure Cooker*, 20 May [retrieved from: http://www.adaptiveaction.org/blog/201505/Escape-the-Pressure-Cooker; accessed 14 April 2016].

Eoyang, G. and Holladay, R. (2013) *Adaptive Action: Leveraging uncertainty in your organization*. Stanford, CA: Stanford University Press.

Meadows, D.H. and Wright, D. (2008) *Thinking in Systems: A primer*. White River Junction, VT: Chelsea Green Publishing.

Miller, J.H. and Page, S.E. (2007) *Complex Adaptive Systems: An introduction to computational models of social life*. Princeton, NJ: Princeton University Press.

Poole, M., Van de Ven, A., Dooley, K. and Holmes, M. (2000) *Organizational Change and Innovation Processes: Theory and methods for research*. Oxford: Oxford University Press.

Rittel, H. and Webber, M. (1973) Dilemmas in general theory of planning, *Policy Sciences*, 4: 155–69.

Zimmerman, B., Lindberg, C. and Plsek, P. (1998) *Edgeware: Lessons from complexity science for health care leaders*. Washington, DC: Plexus Institute.

9 Process mapping: a tool for visualizing system processes from end-to-end

Daniel Cobos Muñoz and Don de Savigny

Introduction

Process mapping and modelling is a systematic approach to understand, analyse, and optimize the processes within complex and adaptive systems in order to achieve intended system goals. A process is a set of activities and tasks that logically group together to accomplish a goal or produce something of value for the benefit of the system and its stakeholder (Davenport and Short, 1990). A process map is a static snapshot, a foundational graphical representation of an end-to-end description of the activities, stakeholders, and requirements of a process.

In health systems, having such a comprehensive and shared view of system processes helps to understand stakeholders' relationships, identify bottlenecks, inefficiencies, and design flaws that limit the performance of the system, and support the integration of new interventions in the system as well as dynamically model, monitor, and manage change over time, once they are implemented. Process mapping is a tool for seeing the whole, and how various stakeholders fit and play their role. It can be transformative in complex systems when managers and major players have 'aha' moments and start to see the system as a whole system. Process mapping can be used for communicating, analysing, sense-making, and managing.

Process mapping and modelling are part of a wider methodology called Enterprise Architecture (EA). EA is a methodology that provides a framework to describe the 'fundamental concepts or properties of a system in its environment embodied in its elements, relationships, and in the principles of its design and evolution' (International Organization for Standardization, 2011: 2). It is a conceptual blueprint that defines the structure and operation of an enterprise. As Zachman pointed out in one of the first papers related to EA: 'The architect's drawings (of a building) are a transcription of the owner's perceptual requirements, a depiction of the final product from the owner's perspective' (1987: 278). The purpose of EA is to bridge the organizing logic of an enterprise (vision and strategy) with its operating model (business processes, information flows and information technology (IT) infrastructure). In a similar way, we can use process mapping to assess whether health system goals and objectives are aligned with its current operations.

EA is a young discipline in the private sector and relatively new for health systems in low- and middle-income countries (LMIC). The Zachman Framework was created in 1987 as a tool to help private organizations to describe how different components and multiple information systems are related in an enterprise

(Zachman, 1987). It was one of the first attempts to rationalize the information system architecture in an organization; and brought methodologies from different disciplines such as building architecture and aeroplane engineering.

Since the 1980s, EA has evolved rapidly with wider approaches to capture not only the information architecture but also the business architecture of organizations. A number of frameworks and methodologies have been developed in the last 20 years to guide the implementation of EA in multiple systems. One of the most widely used is The Open Group Architecture Framework (TOGAF). It is usually described as a cycle and views processes in relation to three 'architecture domains' (see circles A, B and C in Figure 9.1) (The Open Group, 2011). First, the business architecture defines the strategy and describes the key processes of the system. Second, the information system architecture describes the data and information flows in the process as well as the data management activities. Finally, the technology architecture describes all the IT infrastructure and core applications

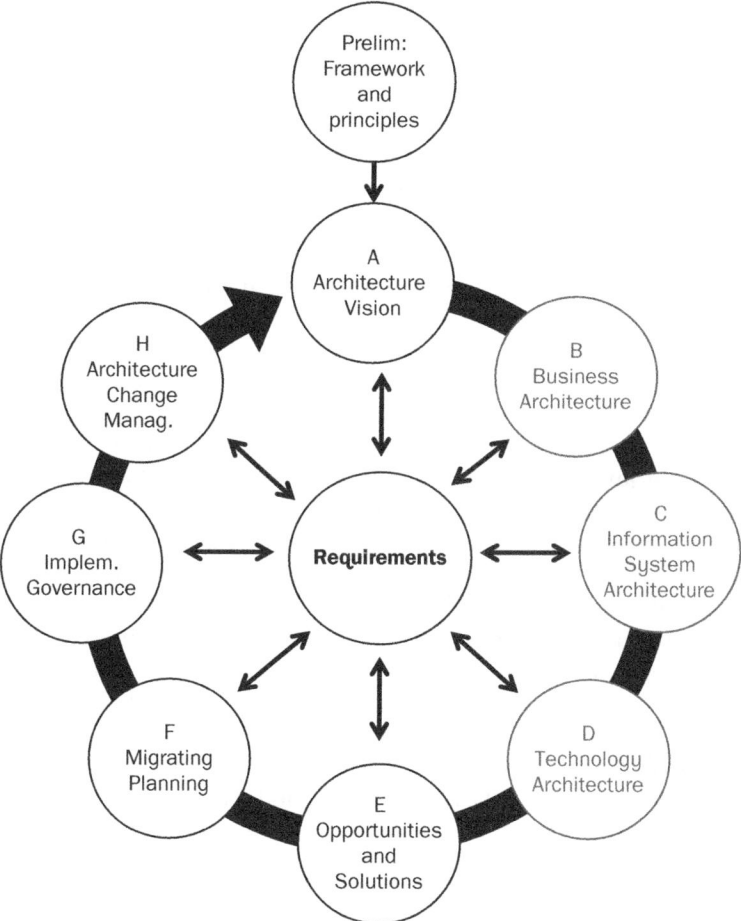

Figure 9.1 TOGAF cycle

Source: adapted from The Open Group (2011).

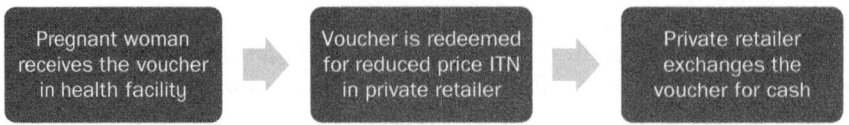

Figure 9.2 Simplified process for a voucher scheme for ITNs

used in each process. In a sense, the three architectures described above are three different representations of the same reality, of the same system.

Health systems, as large complex adaptive enterprises, perform processes to achieve their intended goals and objectives. These processes involve activities implemented by different actors that are structured logically and respond to specific policies and rules. Health systems research and programme implementation have been usually limited by artificial boundaries within the health system (e.g. disease control programmes) and resulted in fragmented information flows and inefficiencies. Process mapping forces researchers and implementers to have a systemic approach, moving across boundaries to describe their processes, and linking them to the overall goals and strategy of the health system.

For instance, countries that aim to reduce the burden of malaria must implement a series of interventions at different levels, such as the provision of adequate malaria prevention and treatment through the distribution of insecticide-treated nets (ITNs); or the implementation of communication campaigns for behaviour change. These interventions entail the effective execution of different activities in a logical way, and are usually performed by different actors in the health system. If we take a voucher scheme for ITNs as an example, a very simplistic way of describing is shown in Figure 9.2.

However, health systems are not as simple as that. There is an extraordinary interplay of issues concerned with governance, human resources, financing, infrastructure, supply chain management, information systems, service delivery, and of course, people as owners and beneficiaries of the product (de Savigny and Adam, 2009). Multiple and competing net delivery systems in the country, fragmented financial sources with foreign and local funds, incentives and disincentives of pregnant women to collect and redeem the vouchers, and the myriad stakeholders involved including public and private sector, were factors that influenced the successful (or not) implementation of the voucher scheme (de Savigny et al., 2012). Process maps can capture some of this complexity in a single diagram (process map) that shows the stakeholders involved in a process and their interactions, responsibilities, and tasks assigned. Figures 9.3 and 9.4 provide an example of one of the processes that would be part on the above-mentioned ITN voucher scheme.

One of the few attempts to use EA to strengthen health systems in LMIC is the Better Immunization Data (BID) Initiative led by PATH, with the objective of empowering countries to 'enhance immunization and overall health service delivery through improved data collection, quality, and use' (BID Initiative, 2014: 2). Using EA, the BID Initiative is trying to bridge the gap between the national immunization comprehensive multi-year plans (cMYP) targets and goals, and the operations of immunization services in a number of countries (e.g. vaccine management, service delivery, vaccine demand generation).

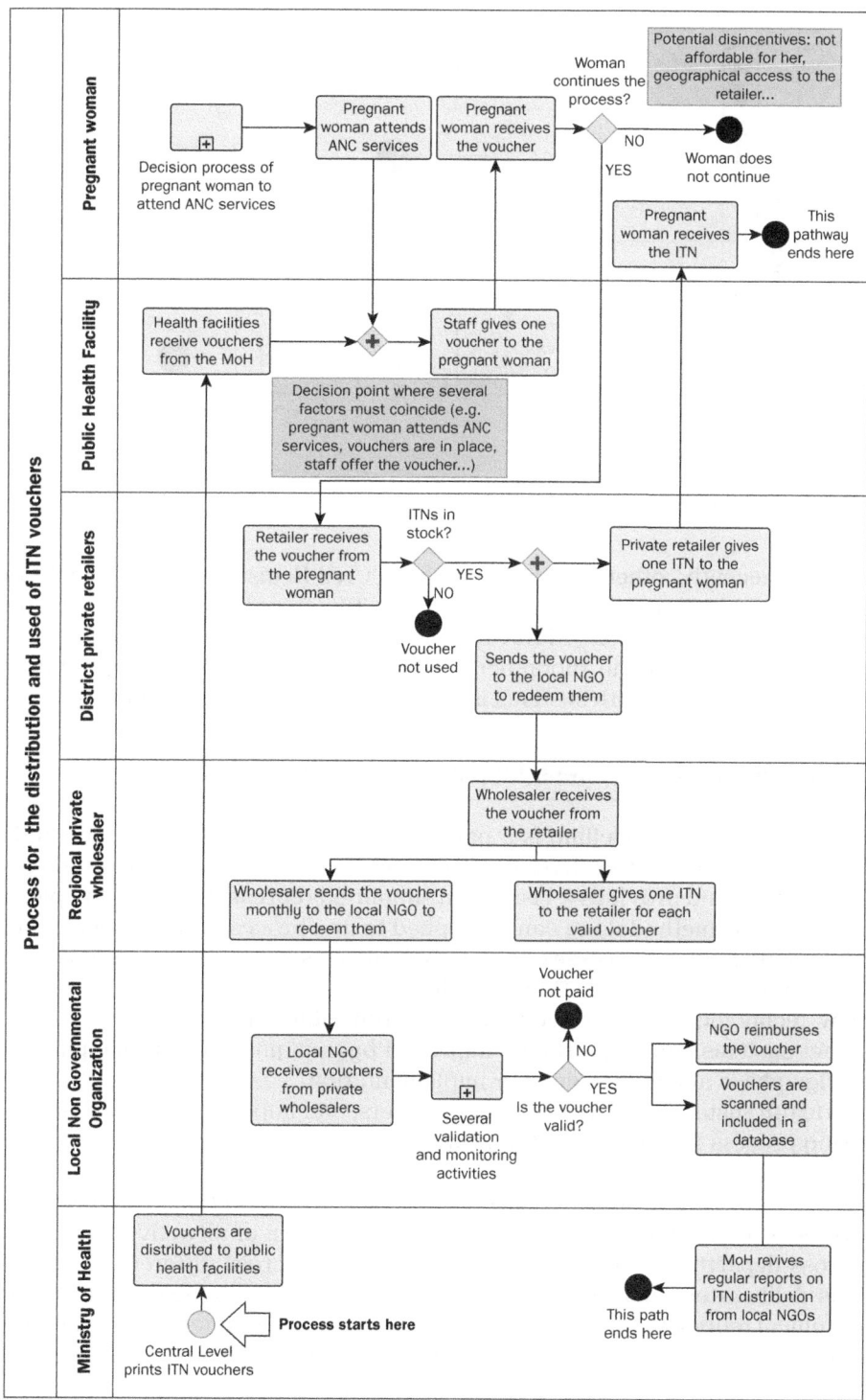

Figure 9.3 Process map for the distribution and use of ITN vouchers in Tanzania

Source: adapted from de Savigny et al. (2012).

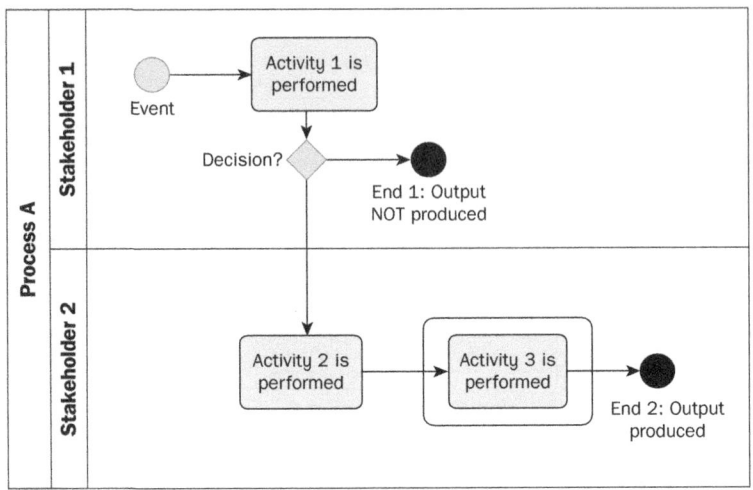

Figure 9.4 Generic process map

More recently, other efforts, such as the Civil Registration and Vital Statistics (CRVS) Digitization Guidebook of the APAI-CRVS from the UN Economic Commission for Africa (UNECA) (APAI CRVS, 2016), and the process modelling activities implemented under the Bloomberg Data for Health Initiative, are applying EA methodologies to health systems in low- and middle-income countries.

Analytic application of process mapping/modelling

Process mapping/modelling is a participatory and action-oriented methodology that promotes collective thinking, facilitates communication among stakeholders in a system, and provides a visual and simplified representation of complex interactions. This methodology can be applied to any process within a health system. Multi-stakeholder processes (e.g. CRVS systems or decentralization), systems with complex information flows (e.g. health insurance or results-based financing), new technologies or services (e.g. mHealth interventions), and digitization of paper systems can be markedly improved by applying process mapping in their design phase and alongside their implementation.

Health system researchers, policy-makers, and implementers can benefit from using process mapping in different ways:

Capture complexity and display health system interactions: Process mapping provides a visually accessible end-to-end description of all activities in a process across departments within an organization, and across different organizations. It is deliberately inclusive, considers all stakeholders involved, and provides a graphical representation of their interactions (or lack of interaction), their responsibilities, and tasks. Process maps serve to document the processes in a system.

Share a common view of the system: Stakeholders usually have different perspectives and understandings of how a system operates. Sometimes they also differ on their vision and objectives for the health system and with abstract constructs difficult to describe. Process mapping offers the opportunity to share these different views in a graphical way and compile them into a single diagram (a process map). Maps are extremely useful to align stakeholders' understanding of how the system operates. They provide a common view of the current system upon which to identify ways to improve process performance.

Model the integration of new interventions: Every new intervention added to a health system will have system-wide effects in the overall system (de Savigny and Adam, 2009). Researchers and policy-makers can use process mapping to model different scenarios for its integration and its potential effects beyond programme boundaries. Stakeholders can hypothesize the effects of interventions using different metrics to assess the performance of a process before implementation.

Stimulate innovative solutions: Process maps are a graphical representation of complex interactions to achieve a goal, and they transcend a system's boundaries. They simplify complex interactions and present them in a graphical format that helps policy-makers and implementers better understand their system as a whole, and set the ground for innovative solutions. Process mapping is a new way of looking at health system processes and stimulates innovative thinking and pioneering solutions that will consider not only the technical aspects of a problem, but also their causal roots and the systemic implications of the solutions.

Manage change: Health systems are dynamic in nature. Every day in any health system, new interventions are implemented and processes reengineered. EA and process mapping offer a structured way to describe where the system is, where it will be and the changes needed to get there in a specific period of time. It includes both the macro view for policy-makers with the vision and strategy of the system, and the micro view for implementers with the detail of stakeholders, responsibilities, and activities in the same tool. It can be used to communicate and advocate for change or to monitor its implementation with time.

Improve compliance with standards: Process maps are a simple, visual, and concise representation of the activities and stakeholders involved in a process. Process mapping can be used to build consensus on how a process must operate and to align stakeholders on the way the process must be implemented.

How to do process mapping and modelling

There are a number of ways to implement process mapping and modelling in a system to address an issue or major objective. We recommend four successive phases. In the first phase, a team with responsibility for overseeing the entire activity is assembled, and all the existing intelligence about the current system's processes for that goal is compiled. In the second phase, the team graphically describes the current end-to-end flow of activities and stakeholders involved in a process using a process map with standard business process mapping notation (BPMN) (As-Is process map). The third phase is an analysis of the As-Is map to identify flaws in the design and areas that could be streamlined to improve the

---

Box 9.1: Phases for the application of process mapping and modelling to health systems

Phase 1 Preparation for the process mapping exercise
  1.1 Assemble the process modelling team
  1.2 Get the necessary support
  1.3 Select the process to improve

Phase 2 Description of the current process
  2.1 Understand the content (Stakeholders map)
  2.2 Complete existing documentation
  2.3 Develop the As-Is process map

Phase 3 Analysis of the current process
  3.1 Assess the performance of the process
  3.2 Root causes of under-performance

Phase 4 Improved process
  4.1 Develop the As-Desired process map
  4.2 Identify the gap and plan for change

---

performance of the whole system. The As-Desired process map is then designed to capture the proposed changes and, finally, the gap between the current and the future situation must be identified (see Box 9.1).

Phase 1: Preparation for the process mapping/modelling exercise

Create a team and get support

The first step in the implementation of process mapping/modelling is to assemble a group who will lead and support the entire exercise. This means that neither will they be responsible for the implementation of the different activities, nor be able to achieve the objectives of the exercise without the support of other members or institutions in the system. The team will be responsible for driving the exercise, and for ensuring that all relevant stakeholders are aligned and involved in the different activities.

As a general rule, the members of the team should be part of the system under analysis, and they will play different roles and have different functions during the exercise. They must have a deep knowledge of the processes and workflows of the system, and they should be appointed and empowered by relevant authorities in the institutions involved in the exercise. Process management and improvement should be a continuous activity in any health system, and having skilled personnel to implement process mapping/modelling as part of the system will ensure continuity and acceptance in the long run. However, the first process mapping exercise will require external support of a process analyst to bring experience

---

Box 9.2: The Business Analyst (BA) is able to analyse the organization and business domain (i.e. CRVS) and document its processes and systems, and assess how technology systems can support the business need

Qualifications, skills, and experience:

- Degree in business or information technology, or similar undergraduate degree
- Business architecture
- Business process modelling notation (UML, Structured, BPMN)
- Systems and data modelling to a logical level, user interface designs
- Functional and non-functional requirements
- Software development life cycle
- Communication skills (oral and written)
- Facilitation skills

Source: CRVS Digitisation Guidebook (APAI CRVS, 2016)

---

and knowledge to the health system. The standard profile of a business analyst can be seen in Box 9.2.

The scale of the process to be reviewed will determine the profile of the members in the core team. For small-scale projects such as improving processes in a district health office, team members will be recruited from the health sector and the exercise confined within the district. In contrast, large-scale system-wide EA process modelling activities will require the participation of several levels of government and will involve several ministries and agencies. Leadership is a critical success factor for the implementation of process modelling in a system.

Select a process to improve

A process is a set of activities and tasks that can logically be grouped together to accomplish a goal or produce something of value for the benefit of the organization, stakeholder or user of the system. A process, then, has the following features:

- It is a set of activities, each of which adds value to the preceding activity.
- The activities are performed by actors that can be either humans or machines.
- It starts with an input (e.g. vital event occurred in a health facility) that is transformed during the process into a valuable output (e.g. national vital statistics).
- The output has value for specific stakeholders.
- It helps accomplish some system's goal.

Processes in health systems can be categorized in three main groups (APAI CRVS, 2016):

1. *Core processes*: those chains of activities by which the system fulfils its mission and achieve its goals.

2. *Support processes*: activities or functions that support the operational functions of the system (e.g. accounting, IT services).
3. *Management processes*: all activities related to planning, decision-making or monitoring the performance of processes within the health system.

The core team must identify and understand all relevant core, support, and management processes that currently exist in the system. Processes are usually interlinked and share some activities. Rather than looking at all problems in all relevant processes in a system, the exercise must focus on a limited number of interlinked processes. The team must decide which processes will be the focus of the exercise in collaboration with other relevant stakeholders. The criteria to select a process or processes are not standard. They must be set up in advance using a transparent mechanism and agreed with all stakeholders. Some examples of selection criteria are:

- Potential of the team to improve it.
- The process is a major cause of low performance of the entire system.
- Cost saving potential.
- Motivation of the staff to change it.
- Complexity of the process.

Phase 2: Description of the current process

The key output from this phase is that all relevant stakeholders understand the process in sufficient detail to enable them to propose meaningful changes in its design. Another output of this phase would be to document the process with all supporting documents that exist (e.g. laws, regulations, standard operating procedures).

Compile existing documentation related to the process

Systems usually have published guidelines, regulations, and standards that provide an overview of the design of the current processes. They represent current design as opposed to current practice. Process mapping is to do with streamlining the design of processes to improve their performance. The fact that processes are not implemented as they were designed could be a consequence of flaws in the design or lack of guidance. How processes are implemented will be a very valuable source of information during the analysis phase.

The team must document not only how the processes are designed but also the system's goals and strategy, and the metrics used to monitor the performance of the system. It will also serve as a resource for advocacy and training once the redesigned process is finally agreed.

Some key documents that could be used as a source of information about health system processes are:

- reports from previous process mapping activities;
- strategic documents with the vision, mission, aims, and goals of the health system or sub-systems;
- standard operating procedures and workflow diagrams;
- operational guidelines, manuals, and protocols;
- job descriptions of staff involved in the process;

Box 9.3: Checklist for completion of Phases 1 and 2

- Describe the end-to-end process in the system.
- Be sure that you have captured all activities and stakeholders involved.
- Balance the effort that it takes to get some information and the benefits of having it.
- Understand the development history of the system but only those facts that are relevant to the current design.
- Focus on how the standard process is rather than capturing its deviations.

- memorandums of understanding between different stakeholders;
- performance monitoring reports;
- international standards with regard to the process under analysis.

It is always difficult to determine when a process has been properly understood and documented. Box 9.3 shows some points that could help the team to make that decision.

Understand the context (stakeholders map)

Systems are usually interlinked and modifications to their processes will have consequences outside of their boundaries. It is important to map the macro environment of any system and to identify all institutions, government bodies, and relevant stakeholders that will be affected by the process modelling exercise.

Stakeholders involved in a process can be broadly categorized into producers, users, and those involved in the governance of the process. Producers are stakeholders directly involved in the implementation of the activities; users are institutions and individuals that benefit from the outputs of the process; and governance entities participate in decision-making, monitoring or governance of the process. For instance, if we take the production of health statistics in a country as an example of a process within a health information system, producers could be considered health facilities or district health offices that generate, compile, and transmit some of the information that goes into the national health statistics. Users would be stakeholders such as the World Health Organization or non-governmental organizations operating in the country. Finally, governance entities in this example could be national committees or specific departments from the Ministry of Health. Note that some stakeholders can be placed in several categories at the same time (e.g. district health offices are producers of health data as well as users of the national health statistics).

At this stage, a more detailed analysis of the stakeholders' environment can be done applying methodologies such as social network analysis described in Chapter 7.

Develop the As-Is process map

The basic tool in process mapping is the process map. The process map is a graphical representation of the flow of activities and the stakeholders involved in a process. It ensures a common understanding on how the process is designed and frames the grounds for the discussions in the analysis stage.

Process maps have standard notation and they follow a specific logic. In 2004, the Business Process Management Initiative (BPMI) developed the first version of a pool of symbols to graphically represent business processes known as Business Process Modelling Notation (BPMN) (Owen and Raj, 2004). BPMN symbols have been used since then as a de facto standard to create process maps and they bridge the gap between designers and stewards of a system and implementers (International Organization for Standardization, 2013). The essential BPMN symbols that health system users require to develop their process maps are shown in Table 9.1.

The first version of the process map is usually based on information in the existing documentation about the process (e.g. SOP, laws, regulations). The first step to develop the As-Is process map is to identify all stakeholders involved in the process and add them as swim lanes in the diagram. The second step will be to identify the different activities involved in the process and the decision points

Table 9.1 Standard BPMN symbols

Symbol	Description
Pool / Lane	Each swim lane represents one stakeholder involved in the process. All activities, decision points or events located in one line are implemented or made by this actor
Task	Represents an atomic activity that cannot be further disaggregated
Start Event	An event that triggers the chain of activities of the process. Usually there is only one start event for one process
End Event	A specific path in the process ends at this point
Interm. Event	An event that occurs within the sequence of activities of a process
⟶	Shows the sequence flow of tasks, gateways, and events in a process
Excl. Gateway	Breaks the flow of activities in two or more mutually exclusive paths based on a condition
Paral. Gateway	Represents two concurrent tasks or paths in a process
Database	Shows where data and information is stored in the process (either paper-based or electronic)
Data object	Represents an exchange of information

that need to be included. Finally, the information flows within the process must be identified and plotted in the map. An example of these three steps applied to a real world example is given in Box 9.4.

There are software solutions to plot process maps using standard BPM notation. These include:

- Bizagi
- Bonita
- Enterprise Architect
- Microsoft Visio.

In addition to describing the flow of activities in a process, the core team must identify the system requirements for this process. System requirements are those critical activities that must be performed within a process to achieve the system goals (see case studies below for an example of system requirements).

'Understanding workshops'

The first draft of the process map based on the documentation available is usually outdated and will contain a number of information gaps. 'Understanding workshops' (concept adapted from Jeston and Nelis, 2014) will provide a forum to refine and enrich the description of the process under analysis. There are two main objectives of these workshops:

1. Obtain all the details to describe the current process from end to end.
2. Bring all relevant stakeholders to a common understanding of how the process is operating under its current design.

These will serve as the foundation for the analysis and redesign phases.

The benefits of workshops as opposed to individual interviews with key informants is that the interactions among the different stakeholders to help achieve a common understanding is as important as obtaining an accurate map of the current process. Sometimes, these workshops will be the first opportunity for all stakeholders to be in the same room discussing a common process in which they are feel involved. Frequently, stakeholders only know those activities in which they are directly involved but they are not aware of their linkages with the wider system.

During the 'understanding workshops' the first draft of the process map must be presented to the group using different means. It could be presented using a laptop and a projector or it might be printed on paper or on posters. Using paper posters adds the advantage of allowing the inputs from the group to be included directly on the map. On the other hand, it could be difficult to work on paper with big groups and using a projector to navigate through the map would be a better option.

The decision of who will participate in the different workshops is crucial for the success of this phase. As a general rule, participants should know the details of the process under analysis; preferably they should be those individuals executing the actual processes (as opposed to managers or policy-makers who think they know the processes). As a group, they should be able to provide the different perspectives from which the process can be described (e.g. regional variations).

Box 9.4: Example of process mapping applied to a health system process

When a **person** dies in a **hospital**, the attending doctor in charge will issue the relatives of the deceased with a medical notification of death detailing the cause of death. The registration of the death in the civil registration office can be done in two ways. For deaths occurring at major hospitals (>200 beds), the **local administrative office** issues a burial permit to the **family** and completes the Death Register Book. The local administrative office also completes an application for a death certificate and sends it to the provincial registrar's office. For a death that occurs in rural areas or small hospitals, the family must take the medical notification of death to the **district registration office** to obtain a burial permit. However, some families decide not to continue the registration process.

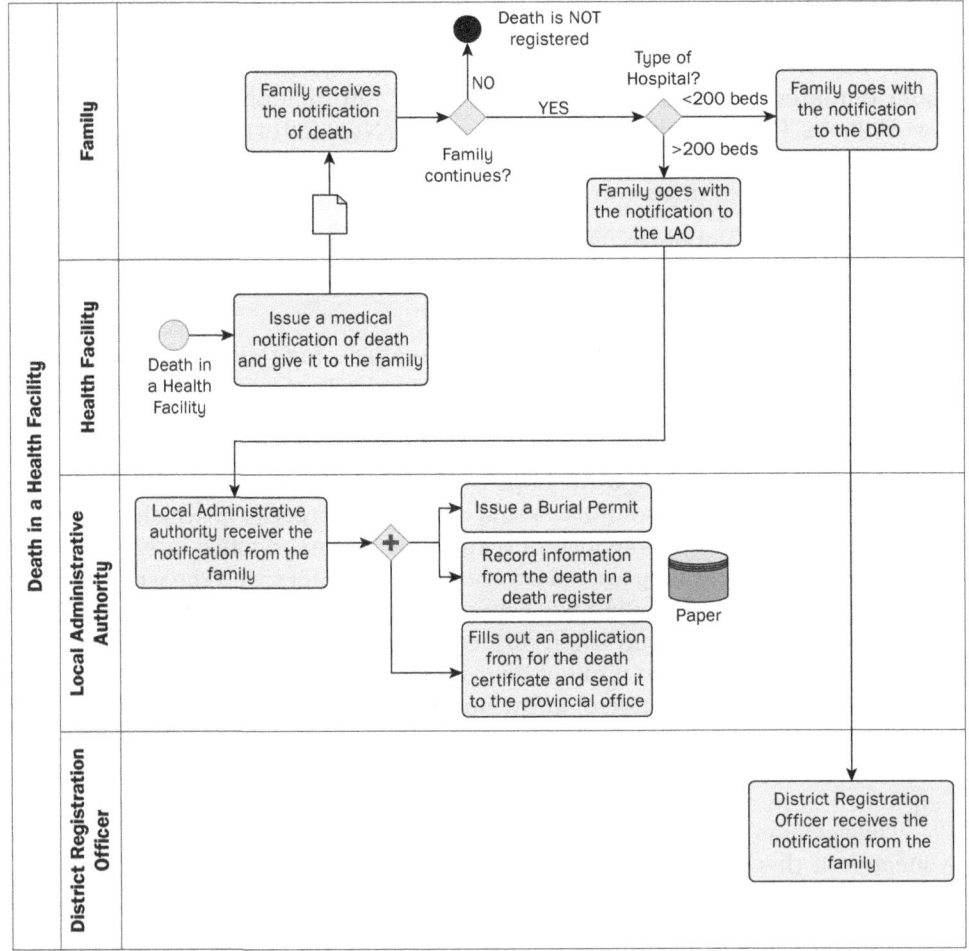

Figure 9.5 Process map of the registration process of a death that occurred in a health facility

It is usually necessary to conduct several understanding workshops to get an accurate picture of the process. There are two main types of 'understanding workshops'. The first type focus on getting all the details needed to refine the first draft of the map. The participants on such a workshop will have a technical profile and should be those persons implementing the process on a day-to-day basis. The other type of workshops focus on getting everyone to a common understanding of how the process is currently designed. These workshops usually include a wider audience, including managers and policy-makers responsible for the performance of the system. In addition, these workshops will explore the macro environment of the system in relation to the process under study and could be used to get the buy-in from the different parts of the system for the exercise.

Phase 3: Analysis of the current process

Once the core team and all relevant stakeholders have a common understanding of how a specific process in the system operates within its current design, the next step is to assess the performance of the process. Based on this assessment, they will identify design flaws and problems that will eventually be improved or redesigned in the As-Desired design.

This phase will answer the following questions:

- Is the process fit for purpose?
- Does it satisfy the business requirements?
- Are there points where the process is delayed or stopped abruptly?
- Are time/resources used efficiently in the process?

Assess the implementation of the process

A series of metrics must be identified to measure the performance of the process under study and understand the root causes of the problems. These metrics will not only show the overall performance of the process and its impact on the whole system, but also the intermediate results and outputs. Thus, they will show if the process is fulfilling the overall goal of the system and they will also be useful to prioritize where the most important problems in the design are. Process metrics will also serve as a baseline that can be compared with future performance once changes in the process have been implemented.

It is also useful at this stage to compare the processes in your system to processes with similar outputs in other systems. Examples of good practices or international standards could be useful to understand the potential of your system in an ideal situation. The performance metrics of the system will provide a transparent way to focus on the key issues that must be considered in the next phase.

The way performance is measured for different processes depends on the type of process and on the overall goals of the system. Some examples of performance measures are given in Table 9.2.

Root cause of under performing

It is not enough to have metrics of how the process is performing within the system to understand the problems of a system. If a process or sub-process is

Table 9.2 Potential performance metrics for health system processes

Example of performance metrics	Source of data
Process and sub-process specific output indicators (i.e. number of ANC visits delivered or number of births registered in the civil registry)	• Routine health information system • Medical records • Surveys • HDSS data • Census
Quality of the outputs delivered by the process	• Complaints register • Existing quality assurance activities
Cost to deliver an output	• Accounting files • Budgets
Human resources required to deliver output/ productivity metrics	• Routine health information system • HR management information system
Number of transactions required to deliver an output	• As-Is map • Interviews with key informants
Process time to deliver an output	• Time–motion surveys
User satisfaction	• Complaints register • Surveys • Interviews with key informants
Existence of bottlenecks	• As-Is map

not performing to acceptable levels, we need to understand why this is happening before proposing any solution. The redesign of a process must address the root causes of the problems and not only its visible effects (i.e. performance metrics).

A good opportunity to identify the root causes of the under-performance of a process is in workshops with technical staff or a wider group of stakeholders. It is difficult to separate the dynamics of understanding the process as it is currently functioning, from the analysis of its problems and design flaws.

Some common issues and design flaws that could exist in health systems are:

• insufficient staff skills to implement the activities;
• non-compliance with the standard process;
• a lack of guidance and standard operating procedures;
• fragmented storage of information;
• parallel channels to transmit the same information;
• weak communication channels among stakeholders involved in the process;
• a high level of bureaucracy;
• insufficient resources to implement the activities.

Methodologies such as causal loop diagrams can be very helpful in this phase (see Chapter 6).

Phase 4: Improved process

The objective of this phase is to redesign the process in such a way as to improve its performance towards fulfilling the goals of the system. There are a few points to bear in mind during this phase:

- The core team will need to consider the process from end to end across all stakeholders involved.
- Any change in the process will have an impact on the rest of the system.
- This phase is not about automating manual activities but rethinking the entire flow.
- Technology is a means to improve the performance and not an end.
- Solutions must be based on the consensus of all relevant stakeholders.

The core team must develop the so-called As-Desired process map and analyse its future implications if it is eventually implemented. The final step in the exercise will be to identify the gap between the As-Is process and the As-Desired process; and plan to move from the current situation to the future design.

Develop the As-Desired process map

Again, one of the best approaches to identify potential improvements to a process is to use a collaborative and participatory workshop with relevant stakeholders. It is essential to provide the participants of these workshops with the right set of objectives for the activity. The workshop should focus on the priority design flaws or problems identified in the previous phases, and which must be coherent with the metrics used for the assessment of the performance of the process.

Participants will also need to agree on the purpose of the redesign exercise, meaning that the group must decide what outputs of the process are most relevant. For instance, it is not the same approach when you are trying to reduce the cost in a process, or to increase coverage of services, or increase user satisfaction. The timeframe to implement the change and alignment with the overall strategy of the system must also be considered from the outset.

During the workshops, participants will brainstorm different solutions and changes in processes that will be based on their own perspective, experience, and interests. It will also be useful to have activities that can help participants look at the process from different perspectives, such as role-plays. Thinking out of the box and finding innovative solutions are essential for the success of this phase.

Most probably, automation of processes will be raised as one of the potential solutions. Transforming manual activities into automatic ones can certainly increase the efficiency and quality of the outputs of the process. However, automation is not an end but a means; the process must drive the technology to be used and not the other way around.

A number of potential solutions or changes to the system will be identified and the core group must assess the future implications of each solution (see Box 9.5). Real-life implementation of the redesigned system using pilot areas provides an

---

Box 9.5: Aspects to consider when analysing future implications of changes in processes

- Legal and regulatory framework in the country
- Human resource requirements
- Job description and potential to assign new functions to existing staff
- Implications of the use of new technology (e.g. maintenance, software modifications)
- Training needs and social capital
- Existing communication channels among stakeholders and potential need for new channels

---

excellent opportunity to test the new process and to understand the feasibility and sustainability of the changes. There are also software solutions (i.e. Bizagi, Bonita, Enterprise Architect) that can model the effects of changes in a process given that we provide the necessary metrics to do it.

Identify the gap and plan for change

With the As-Is and the As-Desired process maps defined, the group must analyse what changes, resources, and new technologies are needed to bridge the gap between the current process and the target system.

There is a large body of literature describing different frameworks and steps to successfully plan for process change. Some elements to consider during this phase are:

- Define governance rules for the entire process and create a governance structures whereby all relevant stakeholders are included.
- Develop communication plan and open communication channels with all relevant stakeholders (including users of the system).
- Design a business case for the target process.
- Draw up a plan of activities and resources needed with clear phases and responsibilities.
- Identify metrics and indicators to monitor change and follow the implementation process.

Practical application case example

Civil Registration and Vital Statistics (CRVS) systems are extraordinarily complex systems and over decades have evolved somewhat differently in various countries. They are sub-systems of the health information sub-system of the health system. But they transcend the health sector, having governance routes in other ministries. Despite the fact that all CRVS systems have the same output objectives, different country CRVS systems have moved along different paths with differing approaches to governance and policies, and differing accountability to multiple ministries such as justice, security, local government or health. Countries also differ in their CRVS organization, implementation, processes, scale, partners, and capacities (WHO, 2013). All CRVS systems are part of their larger political, economic, social, health, and information systems, but nest within them further sub-systems concerned with, for example, legal identity, civil registries, vital statistics, information technologies, and so on. They are the epitome of complex adaptive systems.

Almost all CRVS systems in low- and middle-income countries are failing to achieve adequate levels of coverage and quality despite attempting to apply standard methods proven to work well in high-income countries (Mikkelsen et al., 2015). This suggests system failure rather than technical failure. To date, most attempts to improve CRVS systems have been reductionist and aimed at the technical faults (Abouzahr et al., 2015; Sahay, 2013). As a consequence, they have been slow to achieve results.

Process mapping was applied to CRVS systems as a means to understand its macro environment and to explore the potential uses of this methodology for improving the performance of complex systems. The cases presented here illustrate all the major phases and steps for implementing process mapping and modelling to analyse and redesign complex processes within a health system enterprise.

Case 1 (redesign)

This case study shows how process mapping helped to refine a National CRVS strategy and to define the processes and system requirements needed to fulfil the CRVS goals. Process mapping was used to inform the transition from the current system to the one described in the new strategy using a systems perspective, as opposed to a technology-driven perspective.

The stakeholder environment in the country concerned is highly complex, involving seven government ministries (see Figure 9.6). There is a CRVS ad hoc technical committee but no overarching steering committee concerned with the many agencies doing population registration of one form or another. There is a lack of high-level governance and as a result, some international partners step in, set up and fund their own variations in CRVS, particularly for birth registration, in selected regions of the country, much to the chagrin of the CRVS authorities.

Registration of vital events is mandatory and it is done at the district level using a paper-based system. There is one leading agency for the management and implementation of CRVS activities in the country. The Ministry of Health is responsible for identifying births, deaths, and in particular causes of death, and to report these to the registrar's office. The country completed a CRVS comprehensive assessment in 2014 and the results of this assessment were translated into a strategy starting in 2015 (see Box 9.6).

Implementation of process mapping

The first draft of the maps was developed using the information in the CRVS assessment and other published documents. The first 'understanding workshop'

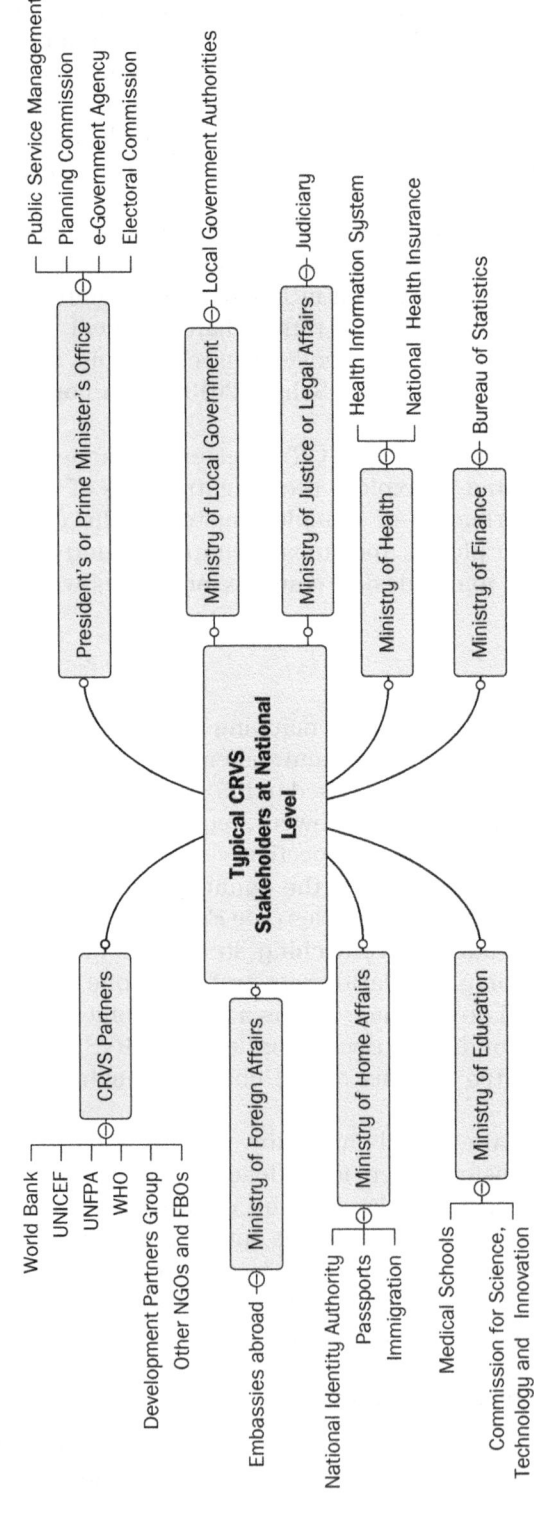

Figure 9.6 Main stakeholders in a typical civil registration and vital statistics system in low- and middle-income countries

---

Box 9.6:   Prominent interventions in the CRVS strategy

- Change the provision of registration services from districts to sub-district wards
- Review and revise the CRVS legal framework
- Use verbal autopsy to ascertain cause of death in the community
- Undertake a system analysis using enterprise architecture and process modeling tools
- Digitize CRVS processes

---

was held at the headquarters of the leading CRVS authority in the country with high-level technical staff from the registrar's agency, the Ministry of Health, and a local health research institute. The contemporary processes for a birth and a death occurring in a health facility and in the community were reviewed and the As-Is process maps were developed. The group also reviewed the National CRVS strategy in the light of the process maps and they decided to work on the As-Desired process map.

One characteristic of all CRVS core processes in the country was that the notification of vital events to the competent authority relied heavily on the family. It is incumbent upon the family to inform the local authority about the vital event. They issue a notification that the family must take to the civil registration office (CRO) to register the event and receive a birth or death certificate. The CRO is located in the district capital and not easily accessible for families living in rural areas. Moving CROs to the ward office or even inside health facilities had already been considered as part of the national CRVS strategy and it would entail a significant change in the existing core processes. The national strategy aimed at having offices located in the wards where the family would be able to register the vital event and obtain the certificate in one interaction with the system.

At this point, the core team realized that the new design had significant implications for the development of the software solution to digitize CRVS processes, it involved multiple stakeholders, and it also implied changes in the actual law that governs CRVS activities. The group decided to organize another 'understanding workshop' with a wider group of stakeholders to present the maps, involve them in the different activities of the process, and push the reactivation of the CRVS steering committee to lead the transition to the target architecture. As a result of this process, the wider stakeholders group decided to:

- Advocate for a High Level CRVS Steering Committee to include the registration authority in the country, National Identification Agency, Ministry of Health, Ministry of Justice, and Ministry of Local Government among other stakeholders under the Prime Minister's Office.
- Conduct a full process mapping/modelling exercise including the enterprise architecture of the CRVS system.
- Pilot the new design in a number of wards before scaling it up.

- Conceptualize how community routine verbal autopsy could be integrated into the CRVS system.
- Slow down the development of the software solution for CRVS digitalization until the system processes have been properly documented and analysed.
- Request a legal review to adapt existing laws and regulations to the new strategy.

Case 2 (engineer new interventions)

This case study illustrates how process modelling can be used to engineer new interventions in an existing system taking into account its impact on a wide variety of stakeholders.

The country in question had a relatively new CRVS system with an established structure to the lowest level of the government administrative system. However, some of the core processes are still under development. It is aimed at producing continuous high coverage real-time vital statistics in a sustainable manner. However, registration coverage was low for births and deaths. The gap between vital events notified by health facilities and registered afterwards was 55% for births and 82% for deaths in 2015.

Implementation of process mapping

For the 'As-Is' situation, the contemporary processes for births and deaths in the community relied on the family to notify and register the vital event. This is particularly relevant for deaths, which primarily occur outside of health facilities. If a death occurs in the community, the family is responsible for notifying the event to a local authority. The local authority will issue a burial permit, report the event to the health facility, and record some administrative information about the deceased in a register kept at the local office. Unlike the country in Case 1 (and many other countries), here almost 100% of deaths have a burial permit issued to the family. Once the family has obtained a burial permit, they must go to the Local Civil Registration Office (LCRO) to register the death and get the death certificate.

One of the major concerns about the performance of the CRVS system was that only 4% of deaths in the country were registered in 2015 because the families were not notifying the deaths, despite almost 100% obtaining burial permits. In addition, for most deaths occurring outside of health facilities, a cause of death was not identified.

The development of the As-Is maps highlighted three existing sub-systems where deaths in the community were captured but they were integrated in the process as a notification step:

- Local administrative authorities were responsible for visiting households in their catchment area on a monthly basis. Part of the information collected during such visits is the occurrence of vital events.
- Community health workers (CHW) have the responsibility, among other things, for family bereavement support.
- CHWs are responsible for reporting vital events in their catchment area.

The core team decided that, in order to increase notification of deaths, they were going to integrate some of the sub-systems already collecting this information. They also envisioned including verbal autopsy using mobile devices as part of their routine CRVS activities for deaths outside of health facilities.

Briefly, in the new processes the local administrative authority will notify electronically health facilities every time they issue a burial permit. They will provide information from the deceased and their family, and the system will generate a unique identifier for the vital event. A new cadre of community health workers will receive notification on a tablet with all the necessary information to locate the family to conduct a verbal autopsy. Once the verbal autopsy has been conducted, information about the cause of death will be redirected to the registrar and the system will use the unique identifier to link this information to the death registration.

The gap analysis identified the need to:

- train community health workers to conduct verbal autopsy with a tablet and develop standard operating procedures;
- acquire IT equipment and set up a maintenance system; and
- design a process to supervise the new system.

References

Abouzahr, C., de Savigny, D., Mikkelsen, L., Setel, P.W., Lozano, R. and Lopez, A.D. (2015) Towards universal civil registration and vital statistics systems: the time is now, *The Lancet*, 386: 1407–18.

APAI CRVS (2016) *CRVS Digitisation Guidebook* [available at: http://www.crvs-dgb.org/en/; accessed September 2016].

BID Initiative (2014) *Product Vision for the Better Immunization Data (BID) Initiative*. Seattle, WA: PATH.

Davenport, T.H. and Short, J.E. (1990) The new industrial engineering: information technology and business process redesign, *Sloan Management Review*, 31 (4): 11–27.

de Savigny, D. and Adam, T. (2009) *Systems Thinking for Health Systems Strengthening*. Geneva: World Health Organization.

de Savigny, D., Webster, J., Agyepong, I.A. (2012) Introducing vouchers for malaria prevention in Ghana and Tanzania: context and adoption of innovation in health systems, *Health Policy and Planning*, 27 (suppl. 4): iv32–iv43.

International Organization for Standardization (ISO) (2011) *Systems and Software Engineering: Architecture description*. ISO/IEC/IEEE4201. Geneva: ISO.

International Organization for Standardization (ISO) (2013) *Information Technology: Object management group business process model and notation*. ISO/IEC 19510. Geneva: ISO.

Jeston, J. and Nelis, J. (2014) *Business Proccess Management: Practical guide to successful implementations*. New York: Routledge.

Mikkelsen, L., Phillips, D.E., Abouzahr, C., Setel, P.W., de Savigny, D., Lozano, R. et al. (2015) A global assessment of civil registration and vital statistics systems: monitoring data quality and progress, *The Lancet*, 386: 1395–1406.

Owen, M. and Raj, J. (2004) BPMN and business process management: an introduction to the new business process modeling standard. *BP Trends*, March: 1–25 [available at: https://pdfs.semanticscholar.org/4522/faedc172d530179ea4d1648c6e5c4b5e42bc.pdf].

Sahay, S. (2013) *Systematic Review of eCRVS and mCRVS Interventions in Low and Middle Income Countries.* Geneva: World Health Organization.

The Open Group (2011) *Open Group Standard. TOGAF Version 9.1.* The Open Group.

World Health Organization (WHO) (2013) *Civil Registration and Vital Statistics 2013: Challenges, best practice and design principles for modern systems.* Geneva: WHO.

Zachman, J.A. (1987) A framework for information systems architecture, *IBM Systems Journal,* 26 (3): 276–92.

10 Systems dynamics: a tool for modelling and testing solutions

Erik Pruyt

Introduction

HIV-AIDS, coronavirus (causing severe acute respiratory syndrome (SARS) in Asia), cholera in Haiti, pandemic H1N1/09 virus across the globe, Ebola virus in West Africa, Zika virus in the Americas: every few years, the world is shaken by the headlines about yet another outbreak or epidemic of a 'new' or 'imported' disease. An underlying catalyst of many such outbreaks and epidemics is the increased interconnectedness of our world.

In terms of infectious diseases, hardly any community or individual human being is sufficiently isolated from the rest not to be impacted by the circumstances and behaviour of others. Assessing the (individual) health risk due to diseases then requires studying the spread of diseases, the ability of the health system[1] to deal with health issues, and the possible effects of targeted health policies. This either requires investigating the effects of the behaviours of thousands to billions of people (and vectors) on the spread of diseases, or the dynamics resulting from the spread of diseases through the population and the response of the health system.

The same is true of non-communicable diseases and conditions. Obesity and burn-out could be seen as unhealthy conditions of individual persons that require interventions on the individual level, or the result of complex system interactions between people with changing individual and cultural preferences, food systems, labour systems, and health systems that first and foremost require systemic interventions, especially in the food and work systems. Systemic interventions in systems or sub-systems are about changing their purpose and scope, the rules and mechanisms by which they are governed and can be accessed, as well as the policies that steer decision-making within these sub-systems, adapting the set of basic and desirable capabilities sub-systems should provide, and the financing and organization required to deliver these capabilities (i.e. the investments in infrastructures and human resources needed to provide basic and desirable system services).

For example, in many countries, health systems (need to) adapt to double societal ageing (i.e. more elderly and more very old elderly who require more care) and 'welfare' effects (i.e. more chronic diseases and health conditions, as well as relatively more health service demands, which may in turn increase life expectancy). This requires long-term capacity planning and policy-making to incentivize/provide investment in nursing homes (close to family dwellings) and hospitals with provision for the elderly, as well as investment in educational facilities for the education of future geriatricians and nurses, policies to motivate (future) health workers, investment for increasing utilization and efficiency (e.g. by means of health robots and apps), as well as rule-making regarding access to expensive cancer treatments and end-of-life treatments, the financing of the health system, the role and influence of health insurers, etc. Proper understanding of the basic mechanisms underlying changes in, and interactions between, health supply and demand is required to do this well.

There are many important dependencies and systemic effects within and between health and other (sub)systems that need to be taken into account when intervening in one or more of the (sub)systems. For example, additional investment in prevention, if effective, results in less need for cure and care than would be the case without the additional investment in prevention, and vice versa. Curing beyond the minimum and addressing the systemic root causes of unhealthy behaviours may in many cases be more beneficial than curing the symptoms – both for the individual and the overall system. Building new long-term rehabilitation centres next to hospitals may, despite the additional investment required, result in system-wide improvements: relatively more fully recovered patients (i.e. fewer 'revolving door patients'/re-admissions); more free capacity in the hospital, which results in greater throughput; reduced costs of rehabilitation per person, etc. Moreover, health (sub)systems also strongly influence other systems, for example a better health system results in a healthier population – less absenteeism, higher labour participation by the elderly, fewer chronically ill people – which is good for the economy, which in turn may (given the right policies and incentives) result in an even healthier population. More focus on proactively improving the overall health system may therefore in the long term be more beneficial for each individual and the system as a whole than dealing with individual health problems in a reactive way.

These examples show that diseases, population health, health-related issues, and health systems are complex, interdependent 'systems' that are dynamic over time. As a consequence, the larger context and a relatively long time horizon need to be considered, and systemic solutions may be required, to improve the system. Although at first sight, studying such issues and systems may seem extremely difficult, in many cases it is not necessary to explicitly represent the behaviour of each individual person or system element in detail to study the system-wide behaviour and the effectiveness of policies for the overall system, and hence, for persons within the system. If this is the case, then systems dynamics can be used to do so. This is also true of many within-person health issues (i.e. health issues within an individual's body that can be described by overall systems equations instead of descriptions of the behaviour of each individual molecule).

Systems dynamics is a method(ology) for modelling and simulating the dynamics of complex systems as a way of better understanding the overall system, of acquiring insight into the link between the underlying structure and behaviour of the system, discovering potential future system dynamics, identifying high-leverage policies, testing their policy robustness, and planning systemic interventions.

Many types of health-related issues can be addressed with systems dynamics, ranging from healthcare delivery systems and products to population health and health economics (Hirsch et al., 2015). Systems dynamics is also used on many different levels, at the operational level of healthcare delivery units as well as at the strategic whole-systems level, and even at the level of the individual body (Lyons and Duggan, 2015).

In the remainder of this chapter, I introduce the systems dynamics modelling and simulation approach, illustrated by a simple hands-on example, and discussed in a broader perspective.

The systems dynamics method[2]

Systems dynamics is an approach to describe, model, simulate, and analyse dynamically complex issues/systems at a relatively high system level. The core systems dynamics approach was proposed and developed by Jay W. Forrester (1958, 1961, 1968, 1971). Today, systems dynamics is applied all over the world in many different domains, including health policy, resource policy, energy policy, environmental management, public policy, innovation policy, international relations, economic policy, business strategy, and operations management.

It is important to note that despite the fact that the word 'system' is used in systems dynamics and in this chapter, systems dynamics models mostly focus on issues, problems or problematic parts of systems, not on entire systems. Note also that systems dynamics is not the only (computational) systems approach to describe, model, simulate, and analyse complex issues. Systems dynamics can be distinguished from other approaches as follows: systems dynamics is a holistic approach, embedded in an iterative and interactive process, in which a small set of building blocks and specific diagramming conventions are used to build simulation models that enable one to simulate (assumed) feedback and accumulation effects, which are often used to search for policies that structurally improve undesirable 'modes of behaviour' (i.e. dynamics). Each of these elements is discussed below.

A systemic endogenous approach

Systems dynamics modelling and analysis is about being able to provide a systemic view of an issue under study and the interconnections with its environment, and simulate and analyse the resulting systems dynamics over time. Systems dynamicists do that by modelling the perceived underlying (material, informational, etc.) structure of largely closed real-world systems and simulating the model over time. 'Largely closed' refers to the choice of broad model boundaries such that 'the causes creating the behaviour of interest lie within the system' (Forrester, 1994: 254). That is, systems dynamics model boundaries are chosen such that all important elements that influence other parts of the system – and also are significantly influenced by other elements of the overall system – are included as endogenous variables (i.e. variables that are embedded in feedback loops). This is often referred to as an 'endogenous systems perspective'. Elements that (could) significantly impact the system but are not sufficiently influenced by other parts of the system are included as exogenous variables. All other elements are omitted in order to focus on the essence of the system.

System structure and building blocks

Systems consist of elements and interdependencies between those elements. Systems are said to be more than the sum of their constitutive elements. Health systems do not just require health workers, or health infrastructure, or medication, or health policies: health system capabilities require the capacities of all these and other elements – and in sufficient amounts and of the right quality. Systems dynamics models are simplified representations of such systems. Issues are dealt with as if they are systems. These models consist of variables, equations within these variables, and causal relations between these variables. Likewise, systems dynamics models are more than a collection of variables. The combination of the structure of a model (especially the feedback loop structure, see below), the functions, and the values used in the model determine the behaviour of the model. In order to fundamentally change the behaviour of a model, one has to change the feedback loops, stock-flow structures, equations, and/or parameter values. It is often assumed that if the model and the real-world system correspond closely, the changes required to fundamentally change the model behaviour would also change the real-world system behaviour. This is one of the reasons why each variable in a systems dynamics model corresponds closely to a real-world counterpart.

Focus on feedback loops

Relations between variables in systems dynamics models only represent direct causal relations that are considered to be sufficiently important to be included by any of the stakeholders. Later, during the simulation stage, it is possible to determine whether causal relations are indeed important or not. Not included from the start are details that are judged by all stakeholders to be unimportant, as well as weak causal or correlational effects. Two or more causal relations between variables that are connected, such that if one follows the causality starting at a particular variable in the loop one eventually returns to that variable, is called a 'feedback loop'. That is, there is a feedback loop if a change in variable A directly causes a change in variable B that directly causes a change in variable C, which in turn directly causes a change in our initial variable A, as in Figure 10.1b. Variable A's behaviour is therefore (partly) caused by its own past behaviour. Such feedback loops give rise to non-linear behaviour, even if all constitutive causal relations are linear.

There are two different types of feedback loop: positive and negative feedback loops. A feedback loop is considered positive or reinforcing if an initial increase in variable A leads, through the feedback effect, to an additional increase in A, or if an initial decrease in A leads to an additional decrease in A. In isolation, positive feedback loops mostly generate reinforcing (i.e. exponentially escalating) behaviour, which could be extremely beneficial or extremely detrimental. The word 'positive' should thus not be confused with its everyday meaning – for example, vicious circles are positive feedback loops that are highly undesirable. A feedback loop is considered negative or balancing if an initial increase in variable A leads, through the feedback effect, to a decrease in A, and if an initial decrease in A leads to an increase in A. In isolation, such feedback loops mostly generate balancing or goal-seeking behaviour. Negative feedback effects are sources of stability as well as resistance to change. Negative feedback may also cause undesirable behaviour – for example, oscillatory behaviour due to a negative feedback loop with a delay.

Feedback loops rarely exist in isolation: they are mostly linked to other feedback loops, and their relative dominance mostly changes over time. Complex system behaviours often arise due to dominance shifts between feedback loops. Computers are used to numerically solve these models, since it is hard – in many cases impossible – to mentally simulate systems dynamics models with more than a few feedback loops.

Diagrammatic conventions

Different types of diagrams are used in systems dynamics. Causal loop diagrams (CLDs) and stock-flow diagrams (SFDs) are almost always used. CLDs display the most important variables, causal relations and feedback loops and their polarities without distinguishing between different types of variables, whereas SFDs display the stock-flow structures without emphasizing the feedback loops. In systems dynamics, CLDs are mainly used in model conceptualization and to communicate the dominant feedback loops and the shifts between them, whereas computational simulation models are mostly constructed using SFD diagrammatic conventions. These diagrams are built according to specific diagrammatic conventions (Lane, 2000a).

The polarity of causal relations is visualized in CLDs by means of link polarities (+ or –). The meanings of 'positive' links and 'negative' links, again, does not correspond to their everyday meanings. A link between two variables A and B is called positive if an increase in A causes B to rise above what it would have been without the influence of the increase in A, or if a decrease in A causes B to fall below what it would have been otherwise. A link between two variables A and B is called negative if an increase in A causes B to fall below the value it would have had otherwise, or if a decrease in A causes B to rise above what it would have been otherwise. The loop polarity of a feedback loop is determined by its net effect: if the net effect of all causal relations in a loop is negative (i.e. in the case of an odd number of negative causal links in a feedback loop), then the entire loop is negative or balancing, otherwise it is positive.

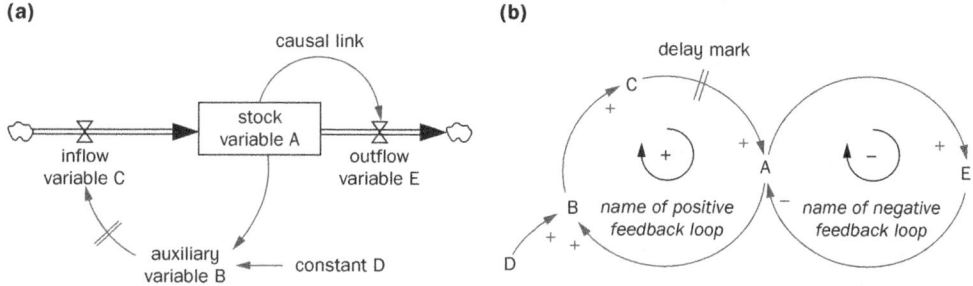

Figure 10.1 Stock-flow diagram (a) and (possible) corresponding causal loop diagram (b)

In SFDs, diagrammatic distinctions are made between four types of variables: stock variables (boxes), flow variables (flows into and out of stock variables, displayed as double arrows with valves), auxiliary variables (no symbol or O), parameters and constants (no symbol or O or ◊). Apart from these variables, there are causal links between variables (single XXXX arrows) and causal links with delay signs (single XXXX arrows with a delay sign). Symbols used in CLDs (link and loop polarities, loop names, etc.) are not included in the corresponding SDFs.

Mathematically speaking, stock variables are integral equations: stock variables accumulate all inflows into the stock minus all outflows out of the stock over time, starting from initial values. In other words, systems dynamics models are essentially systems of integral equations, or systems of differential equations. Note, however, that many systems dynamics models also contain functions that are not found in pure systems of differential equations. Metaphorically speaking, stock variables could be seen as 'bathtubs', inflows as taps or valves, and outflows as drains. Stock variables and flow variables are the essence of systems dynamics models: stock variables, which are often important key performance indicators, can only be changed during simulation by inflows and outflows. Systems dynamics models could in fact be built with stocks and flows and causal links alone. Although auxiliary variables are thus purely auxiliary, in practice systems dynamics models contain many auxiliary variables to keep them understandable and transparent. Lumping all auxiliaries into the flows would result in opaque models with equations that are too complicated to be understood. Flow variables and auxiliary variables often contain special functions. Two types of special functions are extremely important in most systems dynamics models: delay functions to add delay effects and graphical functions to ease the specification of non-linear functions. Apart from these two types of special functions, there are some 20 simple functions that are used regularly (see Pruyt, 2013: chs. 9–11). Besides these regularly used functions, systems dynamics software packages contain hundreds of functions that are rarely used.

Simulation, interpretation in terms of modes of behaviour, and reflection beyond the model

Systems dynamics software packages allow for simulating models over time in such a way that the behaviour of the model unfolds continuously over time. Only the dynamics of well-known and relatively simple systems or extremely small models (up to a few stocks and feedback loops) can be solved analytically or by mental simulation (i.e. without computer simulation). The dynamics of more complex and larger or ill-known systems require numeric integration (i.e. computer simulation). Simulation outcomes always need to be interpreted, since systems dynamics models contain many assumptions, aggregations, simplifications, uncertainties, roughly estimated parameters and initial values. Hence, a model instantiation (i.e. one combination of assumptions, aggregations, simplifications, uncertainties, roughly estimated parameters and initial values), and thus the behaviour generated with the model instantiation, always differ from reality. In systems dynamics, outcomes are never interpreted as point or exact trajectory predictions – they are mostly interpreted as plausible modes of behaviour (i.e. general behaviour patterns). Given the fact that models do not correspond perfectly with the real-world system, there is always a need for extensive sensitivity analysis, scenario analysis, uncertainty analysis, and robustness analysis (i.e. testing whether policies are effective across all uncertainties). These analyses also require human interpretation. Reflection beyond the model is also necessary, since models are always incomplete: despite the fact that models can show side effects, they can only really do so if the corresponding sub-systems, elements, and causal effects that generate or display these side effects are included in the models. Although setting broad boundaries and simulating over long time horizons helps to look at side effects and intertemporal effects, one always needs to reflect consciously about side effects and ethical implications that may not be reflected by the model.

Policies that structurally improve modes of behaviour

Systems dynamics starts from the assumption that the behaviour of a system is largely caused by its own structure. System structure consists of physical and informational aspects as well as the policies and traditions important to the decision-making process (Roberts, 1988). In order to improve undesirable behaviours, it is therefore assumed that the structure of the system needs to be changed. Systems dynamics then allows you to identify desirable system changes and test new policies and their implementation in a 'virtual laboratory' (i.e. by simulating them). Whereas domain experts, decision-makers, and stakeholders can often propose policies based on their knowledge of the issue, experienced modellers can often intuitively distil appropriate structural policies from the structure of a model, from playing with the model, and from performing analyses. There are also advanced analytical methods, statistical screening techniques, techniques to identify dominant loops and shifts in loop dominance, machine learning techniques, and advanced optimization techniques that can be used for advanced policy analysis and policy design.

Iterative and interactive process

Any systems dynamics modelling project consists of different phases or steps. Many systems dynamics modelling projects consist of the following phases or steps: (i) problem identification and problem definition, (ii) system inquiry, (iii) model conceptualization (setting the boundaries, choosing the level of detail, identifying the main variables, qualitatively mapping the relations between the main variables), (iv) model formulation (deciding on the type of variables and adding equations), (v) model testing and fitting, (vi) model simulation, model analysis, and analysis/exploration of the simulation runs, and (vii) policy design, simulation, and testing. Note that systems dynamics modelling is extremely iterative in nature (Homer, 1996). The graph in Figure 10.2, although expressing the iterative character of the systems dynamics process, does not come close to showing how iterative modelling in practice really is. In practice, any of these phases is revisited several times during a systems dynamics modelling project,

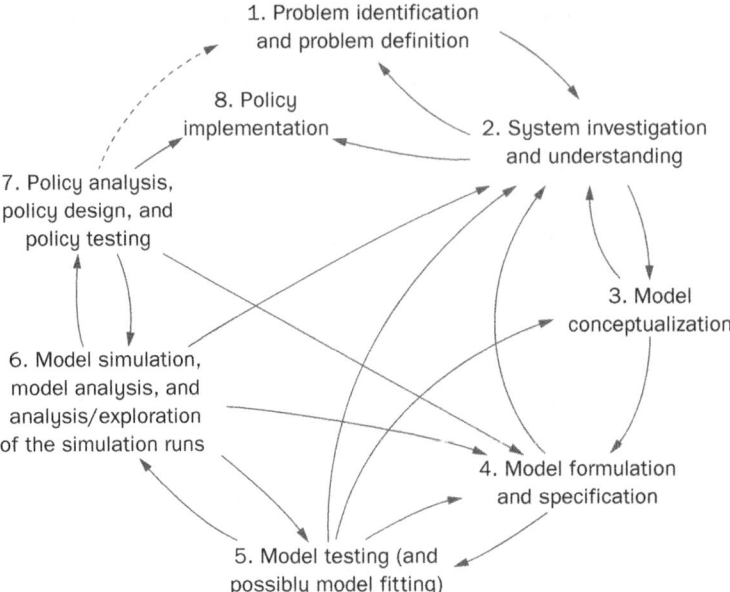

Figure 10.2 The systems dynamics modelling cycle, inspired by Richardson and Pugh (1981) and Forrester (1994)

starting with a small model that is gradually extended until (future) model users or clients have sufficient confidence in the model or in the model-based policy recommendation to use it for the purpose at hand.

Most systems dynamics modelling processes are also highly interactive. That is, a high degree of participation on the part of domain experts, decision-makers, and stakeholders is desirable and often necessary. Modelling often starts with a 'Group Model Building' process (Vennix, 1996) in which mental models of experts, decision-makers, and stakeholders are elicited and joint CLDs are drafted. This qualitative modelling step allows essential qualitative information to be integrated – namely, the mental models about how the system works – into quantitative systems dynamics models. This process mostly takes a few half-day sessions with some 12–20 participants. However, it is also possible to do this in a structured process with more parties/individuals. Afterwards, experts, decision-makers, and stakeholders are closely involved in multiple stages of the process, for a multitude of reasons. A participative/interactive process enables one to (i) exchange and aggregate information, knowledge, and even emotions on existing and desired systems, (ii) gradually develop understanding, policy insights, confidence, and commitment, and (iii) address factors excluded from the actual models (Forrester, 1961, 1971; Lane, 2000b). From a learning point of view, the modelling processes may in many cases be more important than the models and simulation results generated (Forrester, 1985). Jointly constructing a qualitative systems diagram may also be enough to build shared system under-standing, which may in turn be used to build a joint vision. If this is the goal, then it may not always be worth the time and resources to construct a quantitative model.

The systems dynamics method illustrated

The above general explanation of the systems dynamics method is likely to be insufficient for newcomers to start building and simulating their own models. Although it sounds rather too good to be true, after the general theoretical explanation in the previous section, this is exactly what is possible and desirable from this point forward: to build ever more complex models. In the current section, we will therefore build a simple simulation model of a local outbreak of pneumonic plague, starting from scratch. It is strongly recommended that you build your own model and closely follow the description below. Any systems dynamics software package, for example the free Vensim PLE package, can be used. Recommended variable names are in *italics*. This hands-on illustration is based on case 6.8 in Pruyt (2013).

On 3 August 2009, the BBC (2009) reported an outbreak of pneumonic plague in Ziketan, a farming town in the People's Republic of China:

> A second man has died of pneumonic plague in a remote part of north-west China where a town of more than 10,000 people has been sealed off. [T]o prevent the plague [from] spreading, the authorities have sealed off Ziketan, which has some 10,000 residents. About 10 other people inside the town have so far contracted the disease, according to state media. No-one is being allowed [to] leave the area, and the authorities are trying to track down people who had contact with the men who died. [. . .] According to the WHO, pneumonic plague is the most virulent and least common form of plague. It is caused by the same bacteria that occur in bubonic plague – the Black Death that killed an estimated 25 million people in Europe during the Middle Ages. But while bubonic plague is usually transmitted by flea bites and can be treated with antibiotics, [pneumonic plague, which attacks the lungs, can spread from person to person or from animals to people], is easier to contract and if untreated, has a very high case-fatality ratio.

Suppose you need to make a 'quick and dirty' systems dynamics model of this outbreak to test some what-if assumptions and policies: Could this small outbreak become a major outbreak? How fast could that happen? What needs to be done to prevent that from happening?

Open any systems dynamics software package and start a 'new model'. Before building a new model, we need to decide on the time horizon and numeric integration settings to simulate the model. Let us simulate the model over a period of 6 weeks: set the initial time to 0, the final time to 6, the units of time to *weeks*. For the numeric integration method settings, select a small time step (e.g. 0.015625) and an advanced integration method like Runge-Kutta 4 (e.g. *RK4 auto*).

The first step when constructing a systems dynamics model is to identify important stock variables. Building on the well-known SIR model, which stands for **S**usceptible-**I**nfected-**R**ecovered, use the stock variable button to add three stock variables: the *susceptible population*, the *infected population*, and the *recovered population*. You may want to extend the model immediately with an **E**xposed population and develop a SEIR model, which for the sake of simplicity

I postpone until later. The sum of these three stocks is the *total population*. Add the *total population* as an auxiliary variable: this variable simply sums the values of these three stock variables – in itself, it is not an additional integral equation. Add another stock variable, the *cumulative number of pneumonic plague deaths*, to keep track of the casualties due to pneumonic plague.

Flows can now be drawn between these stock variables using the flow equation button, by clicking inside the stock variable from which the flow departs and then clicking inside the stock variable into which the flow variable enters. If a flow equation does not depart from another stock variable in the model, then one needs to click within the canvas first: the flow then departs from outside the model boundary represented by a cloud. Between these stocks, one can draw the following flows: an *infections* flow connecting the susceptible population and the infected population, a *recoveries* flow between the *infected population* variable and the *recovered population* variable, and a *deaths* flow between the *infected population* variable and the *cumulative number of pneumonic plague deaths* variable. The *infections* flow means that when people from the *susceptible population* become infected, they move to the *infected population*.

After connecting stocks and flows, one can use the equation editor to add the equations and initial values of the stock variables. Most packages automatically fill in the integral equations of the stock variables, which means that we only need to set the initial values of the stock variables and fill in appropriate units. Set the initial value of the *infected population* equal to 1, set the initial value of the *recovered population* equal to 0 persons, and set the initial value of the *susceptible population* equal to the *initial total population* minus the *infected population*. The *initial total population* equals 10,000 persons.

Filling in the equations of the flow variables requires adding auxiliary equations and parameters. Assume, for example, that the number of *infections* is equal to the product of the *infection ratio*, the *contact rate*, the *susceptible population*, and the *infected fraction*. Use the auxiliary variable button to add the following auxiliary variables and parameters: the *infection ratio*, the *contact rate*, and the *infected fraction*. Use the causal link button and draw single causal arrows from the *infection ratio*, the *contact rate*, the *susceptible population*, and the *infected fraction* towards the *infections* flow. Use the equation editor to add the equation for the *infections* flow variable. Then proceed by filling in these other variables and parameters. Assume that the average *normal contact rate* amounts to some 15 close contacts per person per week and the *infection ratio* a staggering 75% per contact (i.e. the value 0.75 with units persons), and that the *infected fraction* equals the *infected population* over the *total population*.

If citizens from the *infected population* die, they flow into the *cumulative number of pneumonic plague deaths* stock, else they recover. Model the *deaths* flow as follows: *infected population * fatality ratio/decease time*, and model the *recoveries* simplistically as (1 – *fatality ratio*) – *infected population/recovery time*. Suppose for the sake of simplicity that the average *recovery time* and the average *decease time* are 2 days (i.e. 2/7 of a week). The *fatality ratio* depends on the *antibiotics coverage of the population*, which is 0% at first. Assume the fatality ratio is 90% at 0% *antibiotics coverage of the population* and 15% at 100% *antibiotics coverage of the population*. Although an equation could be used, to add this relation, we could also use a graphical lookup function. Since the *fatality ratio* is – in this first iteration model without policies – constant, it is also possible to simply add the value 0.9. Assume for now, for the sake of simplicity, that those belonging to the *recovered population* do not pose any threat of infection, either because they are really quarantined or because they are not longer contagious.

This concludes the first iteration of systems dynamics model building: the model should look similar to the model in Figure 10.3. It can now be simulated and graphs of the evolution of the sub-populations and other variables can now be displayed (see Figure 10.4).

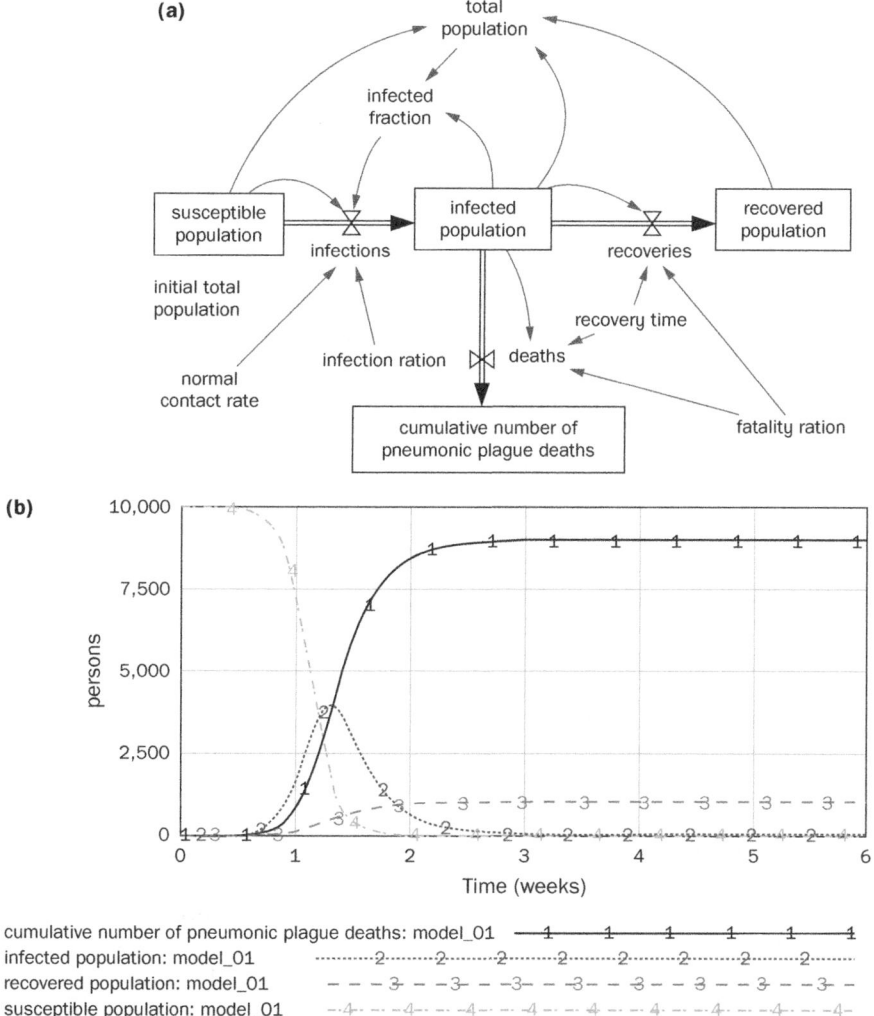

Figure 10.3 Stock-flow diagram of the first iteration model (left) and its behaviour (right) in terms of cumulative deaths (1), infected population (2), recovered population (3), and susceptible population (4)

Note that, given the lack of interventions and change in contact rate in combination with the homogeneous mixing of individuals implicitly assumed in the model, the whole susceptible population would become infected within weeks. These assumptions are unrealistic, however, for outbreaks of lethal contagious diseases such as pneumonic plague are likely to (i) reduce the energy level of those who are infected, resulting in a lower contact rate, (ii) invoke strong feelings of fear and anxiety, which will cause contact rates to drop, and (iii) result in interventions. These effects need to be added in a second iteration model to improve the model for the purposes at hand (i.e. to assess the possible extent and speed of disease spreading and the effectiveness of possible policies).

Let's extend the model with the latter two effects. Assume indeed that the outbreak of pneumonic plague endogenously causes the *contact rate* to drop. Adapt the model by closing the 'loop' between the *infected fraction* and the *contact rate*: create a function *impact of the infected fraction on the contact rate* that, multiplied with the *normal contact rate* (the one without epidemic and panic), gives the effective *contact rate*. Assume that the *impact of the infected fraction on the contact rate* function takes a value of 100% for an *infected fraction* of 0%, a value of 75% for an *infected fraction* of 0.1%, a value of 50% for an *infected fraction* of 1%, a value of 25% for an *infected fraction* of 5%, a value of 12.5% for an *infected fraction* of 10%, a value of 6.25% for an *infected fraction* of 20%, a value of 3.125% for an *infected fraction* of 30%, and so on. Also, suppose that the *antibiotics coverage of the population* increases linearly in the first week of the outbreak from 0% to 100%. Again, we can add a graphical lookup function to simulate this effect. The resulting dynamics are plotted in Figure 10.4.

Although the endogenous reduction of the *contact rate* has a very significant impact on the speed of the outbreak, and the increased availability of antibiotics together with the endogenous reduction of the *contact rate* have an even greater effect on the death toll, the outbreak is not nipped in the bud as should be the case. The reason is that, in this version of the model, the infected population is not quarantined. If, in addition, we add detection, full isolation of the infected population, and full contact tracing after a first diagnosed case (implemented here via the *incubation and isolation time* that drops to virtually nil after two days), then the simulation displayed in Figure 10.5 shows that the outbreak is nipped in the bud for this particular scenario.

Before concluding that this combination of measures is sufficient and robust (i.e. sufficient in all cases), one needs to test the effectiveness of these measures under deep uncertainty (i.e. across all combinations of plausible functions and value ranges). One may also – especially given the small number of infected individuals – want to build and simulate a discrete agent-based model (for a comparison with systems dynamics, see Rahmandad and Sterman, 2008), test the policies across systems dynamics and agent-based models as in Moorlag et al. (2015), or use agent-based models below a threshold and switch to a much more efficient systems dynamics implementation beyond a particular threshold (e.g. in case of vector-borne diseases). One may also want to extend such models, especially in the case of large and lethal yet uncertain outbreaks, with structures that endogenously capture effects of fear and anxiety as in Pruyt et al. (2015) as well as the health system response as in Auping et al. (2016). Figure 10.6 shows part of the

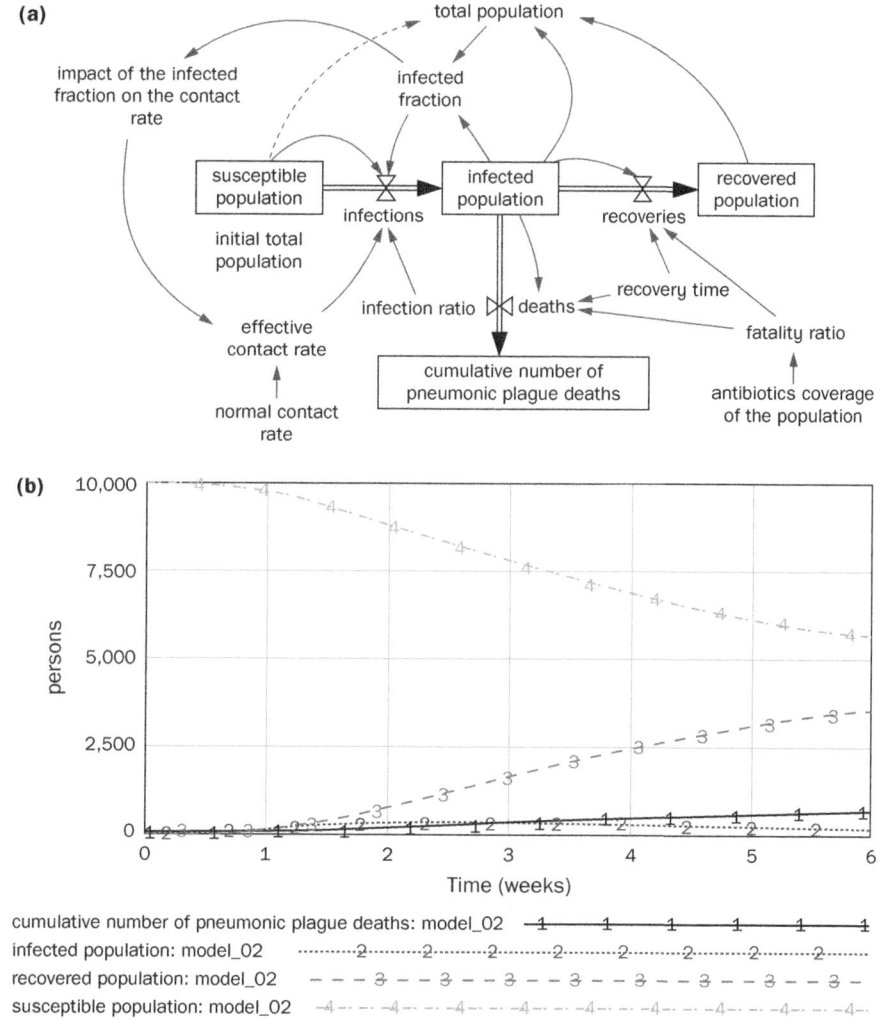

Figure 10.4 Stock-flow diagram of the second iteration model (a) and its behaviour (b)

model structure developed by Auping et al. (2016) to simulate both reactive and proactive capacity policies (in terms of both medical staff and beds) during the 2014 Ebola outbreak in West Africa. The behaviour of the population, the health system, and even the endogenous policy response may have to be included explicitly in order to be able to simulate their interaction effects. Without explicitly modelling the population and health system response, they are likely to overestimate outbreaks.

From simulating diseases to modelling health systems

So far, this chapter has mainly dealt – for ease of explanation – with disease dynamics, not with health systems, which often require bigger simulation models.

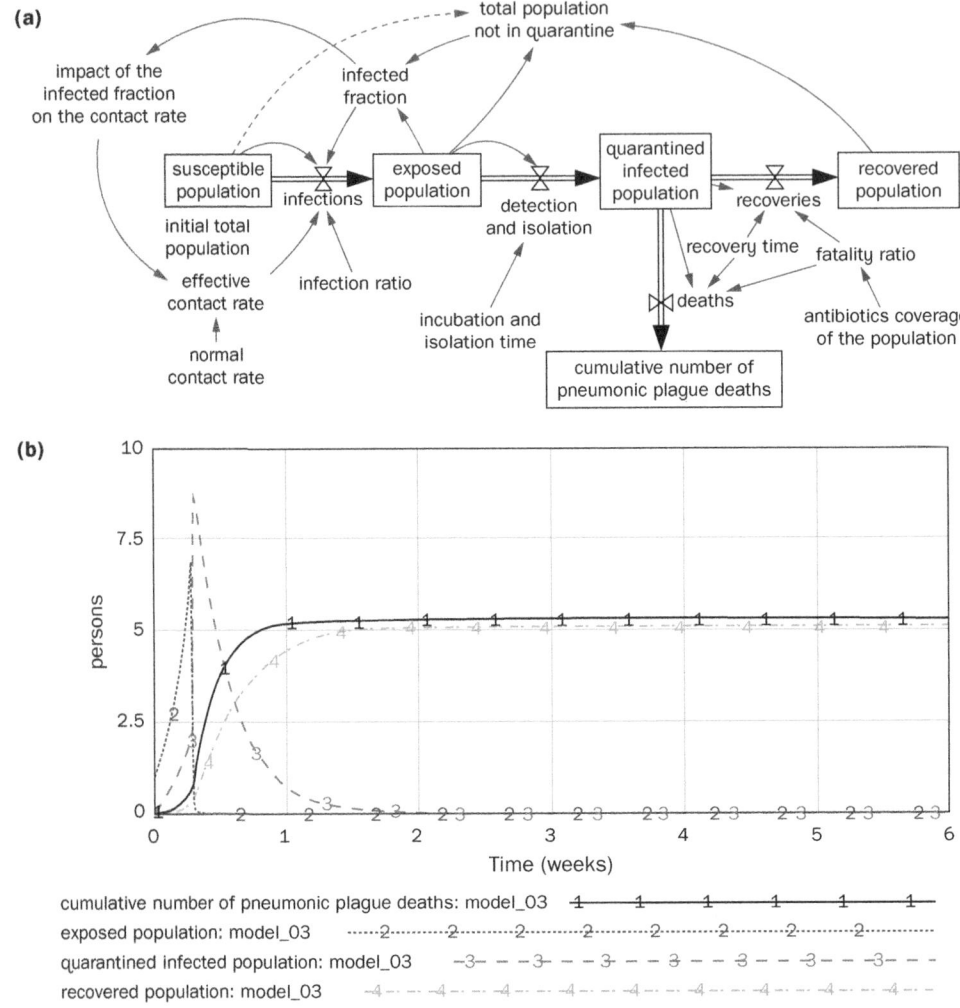

Figure 10.5 Stock-flow diagram of the third iteration model (left) and its behaviour (right)

This may give the impression that systems dynamics is best used for disease dynamics, not for modelling and simulating health systems and health systems policies. However, this is incorrect. Systems dynamics is extremely useful for modelling and simulating the dynamics of health systems. The aim of this section is to illustrate the use of systems dynamics modelling and simulation for health systems without going into too much detail. The first example illustrates the use of systems dynamics to model and simulate the current way in which a particular health (sub)system is organized, and to assess the potential effects of changes to that (sub)system. Figure 10.7 shows a high-level stock-flow representation of the peripheral vascular disease (PVD) 'system'. It shows the different stages of PVD and the transitions between stages are modelled for each age cohort of the population. Worsening PVD symptoms cause potential PVD patients to go to

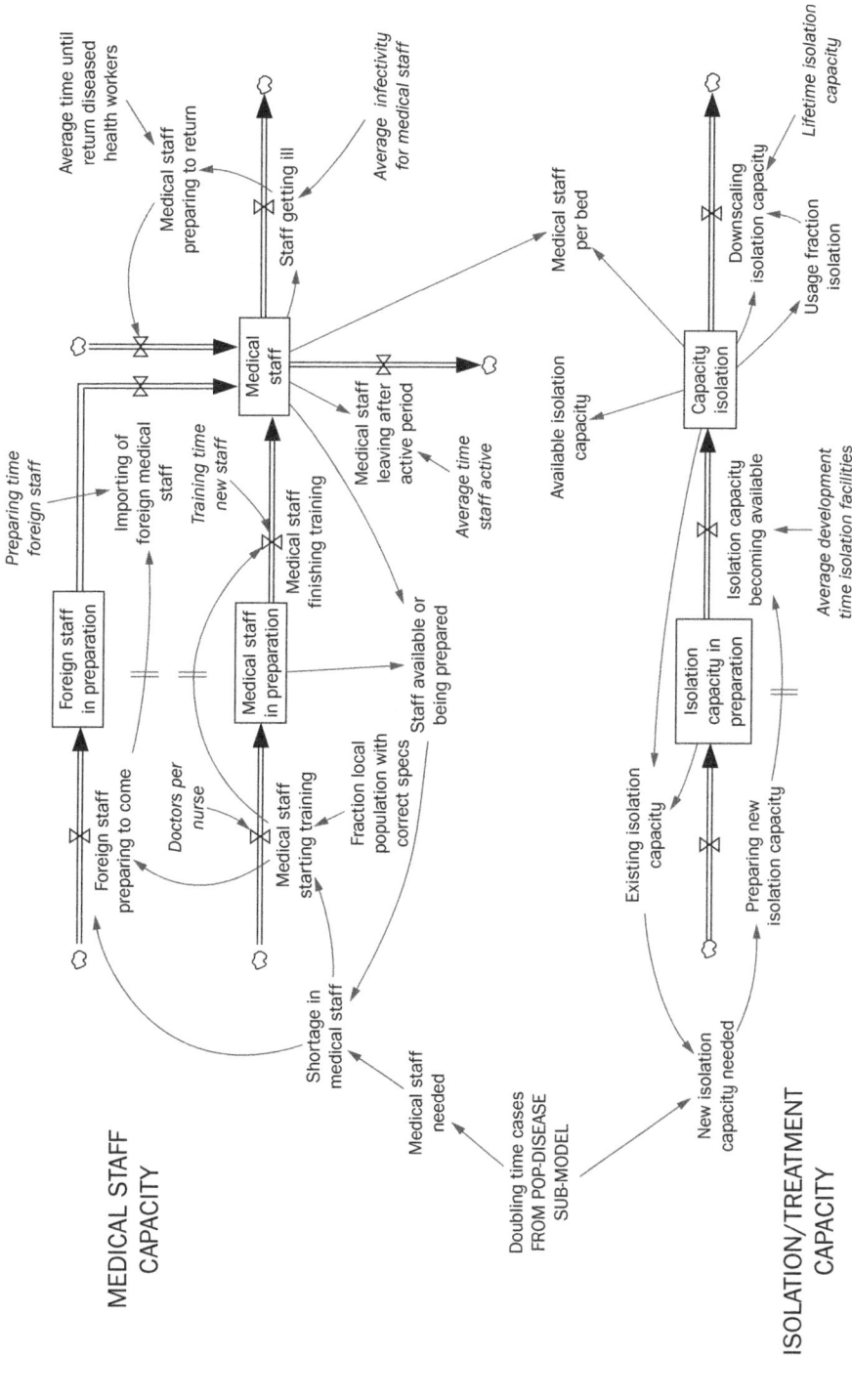

Figure 10.6 Part of the Ebola capabilities sub-model developed and used by Auping et al. (2016)

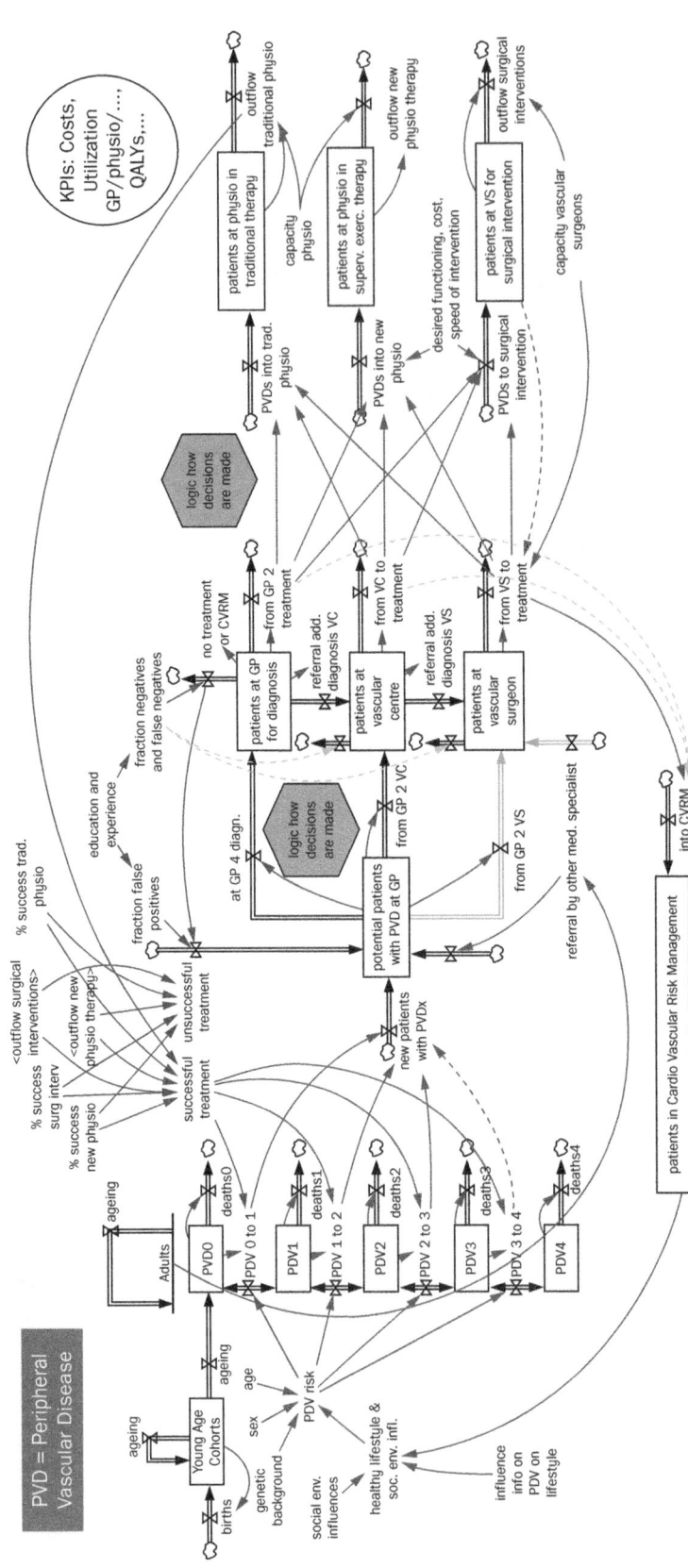

Figure 10.7 High-level model of the peripheral vascular disease (PVD) 'system', including knowledge gaps and potential experiments

their general practitioner (GP). From there on, patients are referred through the health system (to centres that diagnose and to different treatments). When building a model like the one in Figure 10.7, knowledge gaps become visible (i.e. in this case the logic behind referral for diagnosis and the logic behind referral to different treatments). After adapting the model and calibrating it to existing data, one can use it to test what happens, for example, if GPs will be discouraged to refer PDV2 patients to vascular surgeons for diagnosis, and if all patients diagnosed with PDV2 first have to be referred to supervised exercise therapy at the physiotherapist.

Figure 10.8 shows a high-level diagram of a generic systems dynamics model developed for an ongoing system dynamics modelling and simulation project for a health system. The purpose of this modelling project is to identify which investments in cure and care for the elderly need to be made when and where. Using a geo-spatial database, the generic systems dynamics model is parameterized for hundreds of different regions in the country of interest, which after every time step exchange 'migrants' (including mobile elderly seeking a better region to spend their old age), and results are stored to instantiate the model in the next time step. This technique enables one to use systems dynamics simulation models at a low level of aggregation and zoom out to the higher level if desirable, for example, during the policy analysis phase. Using this technique, it is easy to make geo-spatial visualizations and animations. Given the long-term planning horizon associated with societal ageing and the intertia of the health and housing systems,

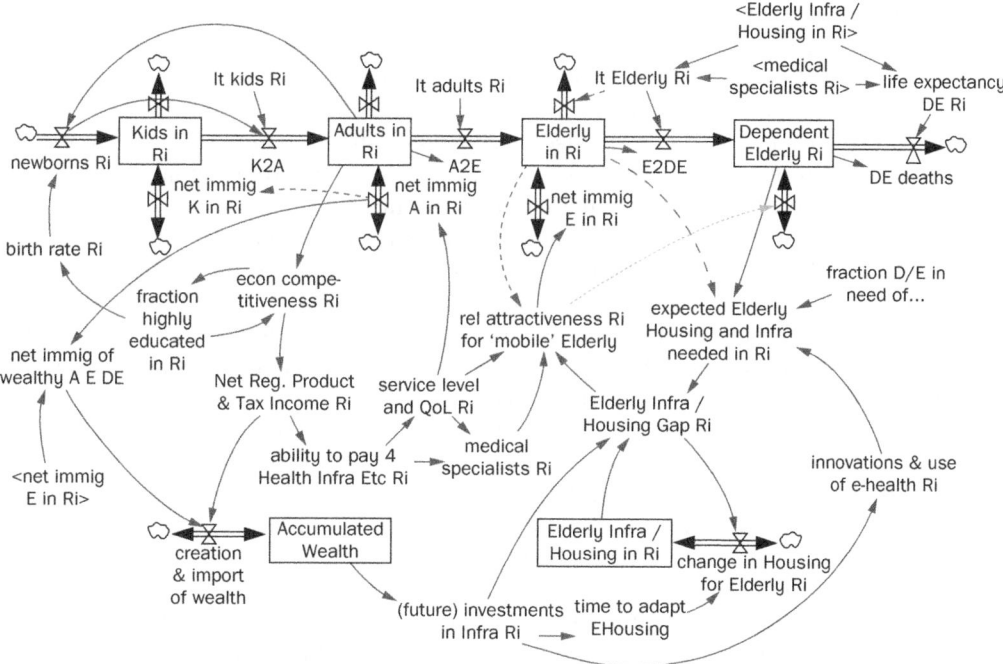

Figure 10.8 Simplified representation of a generic region in a multi-region health systems model

in this case, the model is simulated and the policies (i.e. investment schemes) are tested under deep uncertainty.

How and where to learn systems dynamics (for health-related issues)

Although the core of systems dynamics modelling and simulation is explained and illustrated in the previous sections, it may be useful to consult other sources before starting systems dynamics modelling and simulation. Acquiring basic systems dynamics modelling and simulation skills is relatively easy. There are many introductory systems dynamics textbooks (e.g. Bossel, 2007a; Ford, 2009; Forrester, 1968; Randers, 1980; Sterman, 2000), case books and e-books for hands-on learning (e.g. Bossel, 2007b, 2007c, 2007d; Goodman 1974; Pruyt, 2013), books introducing systems dynamics alongside other computational methods (e.g. Shiflet and Shiflet, 2006), self-study road maps and open online course ware materials, etc. Many of these books and course materials comprise lots of health-related examples and exercises or cases. For example, health-related cases at all levels are covered in Pruyt (2013), and Homer (2012) is fully dedicated to health-related issues. Apart from technical modelling and simulation skills, one needs to acquire observation skills, causal reasoning skills, and conceptualization and aggregation skills, in order to become a system dynamics modeller.

It is more difficult to acquire advanced systems dynamics modelling and simulation skills and to become a professional system dynamics scientist (Sterman, 2002). Some advanced systems dynamics methods and techniques are covered by Rahmandad et al. (2015). Other advanced topics, such as exploratory systems dynamics modelling and analysis (Kwakkel and Pruyt, 2013, 2015; Pruyt and Kwakkel, 2014), adaptive robust policy design (Hamarat et al., 2013, 2014), and behaviour-space scenario generation (Pruyt and Islam, 2015), are discussed in the literature. Advanced topics are also taught in advanced systems dynamics courses offered at universities, at the annual International System Dynamics Summer School, and at the International System Dynamics Conference.[3]

Another fruitful way of learning about systems dynamics applications to health and health systems is the systems dynamics literature on health-related issues. Since systems dynamics is applied extensively to dynamically complex issues in health and health systems, many examples of systems dynamics related work in health and health systems can be found in scientific papers in health-related journals and systems dynamics journals, as well as in the proceedings of the annual International System Dynamics Conference.[4] Good starting points are the special issue of the *System Dynamics Review* on Health and Health Care Dynamics (Dangerfield and Roberts, 1999), and the virtual issue bundling past *System Dynamics Review* articles on health and health systems (Hirsch et al., 2015). Note, however, that many health-related systems dynamics are to be found elsewhere – that is, not in *System Dynamics Review*. The paper by Thompson and Tebbens (2007) on the eradication of polio is an excellent example of impactful systems dynamics work in a health journal. Finally, there are systems dynamics books that focus on aspects of the health systems, such as Paich et al. (2009), which focuses on pharmaceutical market dynamics.

Conclusions

Systems dynamics is a computational systems approach appropriate for modelling and simulating many health-related issues (Peters, 2014). Compared with other approaches, systems dynamics is very accessible, fast in terms of modelling and simulation, and mature. In many cases, it allows one to develop, in a relatively short amount of time, a model that is good enough for the purpose at hand. Note, however, that in other cases, systems dynamics modelling and simulation requires a substantial amount of time and human resources, for example if detailed models are required of very complex systems.

There are many purposes for which systems dynamics modelling and simulation can be used. Although its main purposes are for the joint building of system understanding, generating insights in the link between structure and behaviour, and identifying robust policies, systems dynamics also serves other purposes. Systems dynamics models are used, for example, as engines for flight simulators and serious games.

The high level of aggregation is an advantage as well as a disadvantage. It makes systems dynamics less suitable than detail-focused computational approaches if underlying networks or specific agent-to-agent interactions matter to such an extent that they need to be represented in detail. Note, however, that this may change in the near future with the advent of hybrid approaches (Pruyt, 2015).

Notes

1 The term 'health system' refers here to the organization of means (i.e. people, institutions, protocols, infrastructures, and other resources) to supply healthcare services to meet health demands of a population. In systems dynamics, the word 'system' is used more generally to refer to a 'situation', 'context', 'setting', 'issue' or 'social-technical system' (like the 'health system') – not to a hard, purely technical engineering system.
2 This section is largely based on Pruyt (2013).
3 See http://www.systemdynamics.org/
4 See http://conference.systemdynamics.org/past_conferences/ for SDS conference proceedings since 1976.

References

Auping, W.L., Pruyt, E. and Kwakkel, J.H. (2016) Simulating endogenous dynamics of intervention-capacity deployment: Ebola outbreak in Liberia, *International Journal of Systems Science: Operations and Logistics*, 4 (1): 53–67.

BBC (2009) Plague death toll rises in China. *BBC News* [available at: http://news.bbc.co.uk/2/hi/asia-pacific/8182734.stm; accessed 7 February 2016].

Bossel, H. (2007a) *Systems and Models: Complexity, dynamics, evolution, sustainability.* Norderstedt: Books on Demand GmbH.

Bossel, H. (2007b) *System Zoo 1 Simulation Models: Elementary systems, physics, engineering.* Norderstedt: Books on Demand GmbH.

Bossel, H. (2007c) *System Zoo 2 Simulation Models: Climate, ecosystems, resources.* Norderstedt: Books on Demand GmbH.

Bossel, H. (2007d) *System Zoo 3 Simulation Models: Economy, society, development.* Norderstedt: Books on Demand GmbH.

Dangerfield, B. and Roberts, C. (eds.) (1999) Special issue of *Health and Health Care Dynamics*, 15 (3).

Ford, A. (2009) *Modeling the Environment,* 2nd edn. Washington, DC: Island Press.

Forrester, J.W. (1958) Industrial dynamics: a major breakthrough for decision makers, *Harvard Business Review*, July/August: 37–66.

Forrester, J.W. (1961) *Industrial Dynamics.* Cambridge, MA: MIT Press.

Forrester, J.W. (1968) *Principles of Systems.* Cambridge, MA: Wright-Allen Press.

Forrester, J.W. (1969) *Urban Dynamics.* Cambridge, MA: Productivity Press.

Forrester, J.W. (1971) *World Dynamics.* Cambridge, MA: Wright-Allen Press.

Forrester, J.W. (1985) 'The' model versus a modeling 'process', *System Dynamics Review*, 1 (1): 133–4.

Forrester, J.W. (1994) System dynamics, systems thinking, and soft OR, *System Dynamics Review*, 10 (2/3): 245–56.

Goodman, M. (1974) *Study Notes in System Dynamics.* Cambridge, MA: Wright-Allen Press.

Hamarat, C., Kwakkel, J.H. and Pruyt, E. (2013) Adaptive robust design under deep uncertainty, *Technological Forecasting and Social Change*, 80 (3): 408–18.

Hamarat, C., Kwakkel, J.H., Pruyt, E. and Loonen, E.T. (2014) An exploratory approach for adaptive policymaking by using multi-objective robust optimization, *Simulation Modelling Practice and Theory*, 46: 25–39.

Hirsch, G., Homer, J. and Tomoaia-Cotisel, A. (eds.) (2015) System dynamics applications to health and health care. Virtual issue of *System Dynamics Review*.

Homer, J. (1996) Why we iterate: scientific modeling in theory and practice, *System Dynamics Review*, 12 (1): 1–19.

Homer, J. (2012) *Models That Matter: Selected writings on system dynamics 1985–2010.* Barrytown, NY: Grapeseed Press.

Kwakkel, J.H. and Pruyt, E. (2013) Exploratory modeling and analysis, an approach for model-based foresight under deep uncertainty, *Technological Forecasting and Social Change*, 80 (3): 419–31.

Kwakkel, J.H. and Pruyt, E. (2015) Using system dynamics for grand challenges: the ESDMA approach, *Systems Research and Behavioral Science*, 32 (3): 358–75.

Lane, D. (2000a) Diagramming conventions in system dynamics, *Journal of the Operational Research Society*, 51 (2): 241–5.

Lane, D. (2000b) Should system dynamics be described as a 'hard' or 'deterministic' systems approach?, *Systems Research and Behavioral Science*, 17 (1): 3–22.

Lyons, G.J. and Duggan, J. (2015) System dynamics modelling to support policy analysis for sustainable health care, *Journal of Simulation*, 9 (2): 129–39.

Moorlag, R., Pruyt, E., Auping, W.L. and Kwakkel, J.H. (2015) Shale gas and import dependency: a multi-method model-based exploration, *International Journal of System Dynamics Applications*, 4 (1): 31–56.

Paich, M., Peck, C. and Valant, J. (2009) *Pharmaceutical Product Branding Strategies: Simulating patient flow and portfolio dynamics,* 2nd edn. New York: Informa Healthcare.

Peters, D.H. (2014) The application of systems thinking in health: why use systems thinking?, *Health Research Policy and Systems*, 12 (1): 51 [doi: 10.1186/1478-4505-12-51].

Pruyt, E. (2013) *Small System Dynamics Models for Big Issues: Triple jump towards real-world complexity.* Delft: TU Delft Library [this e-book can be downloaded for free from: http://simulation.tbm.tudelft.nl/smallSDmodels/Intro.html].

Pruyt, E. (2015) From building a model to adaptive robust decision making using systems modeling, in M. Janssen, M.A. Wimmer and A. Deljoo (eds.) *Policy Practice and Digital Science: Integrating complex systems, social simulation and public administration in policy research.* Public Administration and Information Technology, Vol. 10 (pp. 75–93). Dordrecht: Springer.

Pruyt, E. (2016) Integrating systems modelling and data science: the joint future of simulation and 'big data' science, *International Journal of System Dynamics Applications*, 5 (1): 1–16.

Pruyt, E. and Islam, T. (2015) On generating and exploring the behavior space of complex models, *System Dynamics Review*, 31 (4): 220–49.

Pruyt, E. and Kwakkel, J.H. (2014) Radicalization under deep uncertainty: a multi-model exploration of activism, extremism, and terrorism, *System Dynamics Review*, 30 (1/2): 1–28.

Pruyt, E., Auping, W.L. and Kwakkel, J.H. (2015) Ebola in West Africa: model-based exploration of social psychological effects and interventions, *Systems Research and Behavioral Science*, 32 (1): 2–14.

Rahmandad, H. and Sterman, J.D. (2008). Heterogeneity and network structure in the dynamics of diffusion: comparing agent-based and differential equation models, *Management Science*, 54 (5): 998–1014.

Rahmandad, H., Oliva, R. and Osgood, N.A. (eds.) (2015) *Analytical Methods for Dynamic Modelers*. Cambridge, MA: MIT Press.

Randers, J. (1980) *Elements of the System Dynamics Method*. Cambridge, MA: MIT Press.

Richardson, G.P. and Pugh, A.I. (1981) *Introduction to System Dynamics Modeling*. Portland, OR: Productivity Press.

Roberts, E. (1988) *Managerial Applications of System Dynamics*. Cambridge, MA: MIT Press.

Shiflet, A. and Shiflet, G. (2006) *Introduction to Computational Science: Modeling and simulation for the sciences*. Princeton, NJ: Princeton University Press.

Sterman, J.D. (2000) *Business Dynamics: Systems thinking and modeling for a complex world*. Boston, MA: Irwin/McGraw-Hill.

Sterman, J.D. (2002) All models are wrong: reflections on becoming a systems scientist. *System Dynamics Review*, 18 (4): 501–31.

Thompson, K. and Tebbens, R.J. (2007) Eradication versus control for poliomyelitis: an economic analysis, *The Lancet*, 369 (9570): 1363–71.

Vennix, J.A.M. (1996) *Group Model Building: Facilitating team learning using system dynamics*. New York: Wiley.

Appendix

Model version 1: equations

contact rate = 15 ~ persons/persons/Week

cumulative number of pneumonic plague deaths = INTEG (deaths , 0) ~ persons

deaths = infected population * fatality ratio / recovery time ~ persons/Week

fatality ratio = 0.9 ~ Dmnl

infected fraction = infected population / total population ~ Dmnl

infected population = INTEG (infections - recoveries - deaths , 1) ~ persons

infection ratio = 0.75 ~ persons/persons

infections = contact rate * infection ratio * susceptible population * infected fraction ~ persons/Week

initial total population = 10000 ~ persons

recovered population = INTEG (recoveries , 0) ~ persons

recoveries = (1 - fatality ratio) * infected population / recovery time ~ persons/Week

recovery time = 0.285714 ~ Week

susceptible population = INTEG (– infections, initial total population – infected population) ~ persons

total population = infected population + recovered population + susceptible population ~ persons

FINAL TIME = 6 ~ Week

TIME STEP = 0.015625 ~ Week

Model version 2: changed and added equations

antibiotics coverage of the population = WITH LOOKUP (Time/timevar, ([(0,0)-(10,10)], (0,0),(1,1),(10,1))) ~ Dmnl ~ timevar is used to (ab)use a withlookup function as timeseries variable

effective contact rate = normal contact rate * impact of the infected fraction on the contact rate ~ persons/persons/Week

fatality ratio= WITH LOOKUP (antibiotics coverage of the population, ([(0,0)-(1,1)],(0,0.9),(1,0.15))) ~ Dmnl

impact of the infected fraction on the contact rate= WITH LOOKUP (infected fraction, ([(0,0)-(1,1)],

(0,1),(0.001,0.75),(0.01,0.5),(0.05,0.25),(0.1,0.125),(0.2,0.0625),(0.3,0.0325),(1,0))) ~ Dmnl

infections = effective contact rate*infection ratio*susceptible population*infected fraction ~ persons/Week

normal contact rate= 15 ~ persons/persons/Week

timevar=1 ~ Week ~ timevar is used to (ab)use a withlookup function as timeseries variable

Model version 3: changed and added equations

deaths = quarantined infected population*fatality ratio/recovery time ~ persons/Week

detection and isolation = exposed population/incubation and isolation time ~ persons/Week

exposed population= INTEG (infections-detection and isolation, 1) ~ persons

incubation and isolation time= 0.285714 -STEP(0.28, 0.285714) ~ Week

infected fraction=exposed population/total population not in quarantine ~ Dmnl

quarantined infected population= INTEG (detection and isolation-deaths-recoveries, 0) ~ persons

recoveries= (1-fatality ratio)*quarantined infected population/recovery time ~ persons/Week

susceptible population= INTEG (-infections, initial total population-exposed population) ~ persons

total population not in quarantine = exposed population + recovered population + susceptible population ~ persons

Scenario technique: a tool for simulating and reflecting on alternative solutions

Horst Christian Vollmar

It is less important to foresee the future than to be prepared for it.

Perikles, 493 – 429 v. Chr.

Introduction

Over the last decades, the scenario method has become an important tool in fore-sight activities and research. Scenario development emerged after the Second World War in military strategic planning (van Notten, 2006). The term 'scenario' (from the Latin word *scaena*) means 'scene' and was originally used in the context of such performing arts as theatre and film. Herman Kahn originally adopted the term and introduced the concept of scenario development during his time at the RAND Corporation (Kahn and Wiener, 1967). In the 1960s, Royal Dutch Shell and General Electric introduced scenario techniques in their corporate planning procedures (van Notten, 2006). After a temporary decline in the 1980s, the scenario technique has become more commonly used for strategic planning (Schwartz, 1996; Varum and Melo, 2010).

There are varying definitions of 'scenario', or in Glenn's words: 'scenario is the most abused term in futures research' (Glenn and The Futures Group International, 2011: 2). However, there is one point upon which there is consensus: scenarios are not predictions (Van der Heijden et al., 2002). Because there are so many definitions and various typologies (Bishop et al., 2007), this chapter cannot be considered general guidance to develop scenarios (see recommendations for further reading). Glenn gave the following definition: 'a scenario is a story with plausible cause and effect links that connects a future condition with the present, while illustrating key decisions, events, and consequences throughout the narrative' (Glenn and The Futures Group International, 2011: 3). Another definition that covers many of the characteristics proposed by others is: 'Scenarios are consistent and coherent descriptions of alternative hypothetical futures that reflect different perspectives on past, present, and future developments, which can serve as a basis for action' (van Notten, 2006: 2). Characteristics inherent in the various definitions include that they are: causally coherent, internally consistent, hypothetical, and/or descriptive (van Notten, 2006). However, scenarios are always hypothetical, they are not arbitrary (Kosow and Gaßner, 2008).

The creation of scenarios presents an interdisciplinary approach to explore future issues while offering several advantages – for example, the support of a future-oriented way of thinking by taking alternative developments into consideration (Glenn and The Futures Group International, 2011; Kosow and Gaßner, 2008; Schwartz, 1996). Furthermore, it fosters systematic and structured discussion of uncertain alternative futures through the incorporation of expert knowledge. Proceeding step-by-step reduces the perceived complexity of the correlations examined, generates findings that are comprehensible (Wilkinson et al., 2013), and should improve strategic decision-making (Gregorio et al., 2014; Moore et al., 2007; Postma and Liebl, 2005). It may be combined with other foresight methods such as the Delphi technique or road mapping (Anderson and McConnell, 2007; Gnatzy and Moser, 2012). Although less commonly used in the context of health and health care than either the Delphi technique or simulation modelling, the scenario method has also been used to support strategic decision-making in the field (Vollmar et al., 2015).

According to Kosow and Gaßner (2008), there are three major categories of scenario development methods:

1. Creative-narrative scenario techniques
2. Scenarios based on trend extrapolation
3. Systematic-formalized scenario techniques.

Given this is only one categorization in an area with several classification systems, it will be useful to explain the scenario technique. The examples in this chapter follow the above categorization (and an example of each of 1 and 3 is provided).

Limitations of the scenario technique

The following limitations inherent to the scenario technique should be borne in mind (Amer et al., 2013; Glenn and The Futures Group International, 2011; Kosow and Gaßner, 2008; Lexa and Chan, 2010; Schreuder, 1995). First, creating explorative scenarios can be time-consuming and therefore cost-intensive, in particular because they tie up personnel resources (Vollmar et al., 2014). However, the processes are scalable; a small group (of students, for example) might be able to develop consistent scenarios (Albers and Broux, 1999; Bezold, 1992; Rydstrom and Tornberg, 2006). Second, the quality of the scenarios depends greatly on the imagination, information basis, and competency of the experts taking part (Baker et al., 2006; Vollmar et al., 2014). Thus, there is a potential risk of biased scenarios if experts are inclined to prefer well-known developments and to reject any that seem too unorthodox; in other words, opinion leaders who try to dominate a scenario group are counterproductive (Retel et al., 2012). As a result, the selection of experts is of considerable importance and should be based on the criteria applied to consensus processes and Delphi methods (Baker et al., 2006; Retel et al., 2012). However, it is possible to obtain usable conclusions with 'ordinary' persons acting as experts (Karger, 2013; Nguyen et al., 2014; Niewoehner et al., 2005). Third, if the scenario development process is not only narrative, but also includes quantitative aspects by means of calculations (Becker, 1988; Retel et al., 2012), the mathematical processes used to generate the scenarios may be plausible for scenario-natives but incomprehensible for non-experts (Vollmar et al., 2014). Fourth, the selection of the drivers or key factors (or the preliminary considerations) is the crucial point of each scenario analysis (Glenn and The Futures Group International, 2011). In principle, a systematic search for each selected key factor is desirable to generate evidence, but available resources would hardly allow this. Fifth, scenarios do not forecast the future, as each step always entails subjective assessments and evaluations of abstract and complex facts. So, another threat might be the overestimation of the exactness of explorative scenarios. The sixth and final point seems to be critical for the acceptance of the scenario method as a scientific tool, particularly in the field of public health. Even though the reporting of the more common Delphi technique has room for further improvement (Diamond et al., 2014; Landeta, 2006), it seems that the variability in using and reporting the scenario method is much higher (Amer et al., 2013; Gerhold et al., 2015; Kosow and Gaßner, 2008; Postma and Liebl, 2005). There is a need to improve the reporting of scenario projects, along the lines of a GRAMMS-like guideline that is used for mixed methods studies and recommended by the equator-network (www.equator-network.org) (O'Cathain et al., 2008).

Overview of using the scenario technique in the healthcare context

A recent scoping review identified a total of 41 journal articles on using the scenario technique in the context of health and health care (Vollmar et al., 2015). These 41 publications described 38 different scenario projects in total.

The focus of the scenario projects differed in many ways. Most of them address disease-related issues ($n = 9$), in particular mental health and dementia ($n = 4$) (Bierbooms et al., 2011; Bijl, 1992; Bijl and Ketting, 1991; Karger, 2013; Vollmar et al., 2014) and cancer ($n = 3$) (Retel et al., 2012; Rydstrom and Tornberg, 2006; Schaapveld and Cleton, 1989). Five scenario projects deal with public health issues on an organizational level (Bezold, 1992; Neiner et al., 2004; Nguyen et al., 2014; Venable et al., 1993; Zentner, 1991) and five on the labour market of different healthcare professionals (Buchan and Seccombe, 2012; Eberl and Schnepp, 2006; Gregorio et al., 2014; Leufkens et al., 1997; Rhea and Bettles, 2012), with two of them focusing on the pharmacy profession (Gregorio et al., 2014; Leufkens et al., 1997). In addition, four projects address health care 'in general' (Bezold, 2005; Ling and Hadridge, 2000; Ma and Seid, 2006; Nielsen, 1996), four projects other technology developments (Enzmann et al., 2011; Meristo et al., 2009; Retel et al., 2012; Wiek et al., 2009), and an additional four the field of biotechnology and personalized medicine (Karger, 2013; Niewoehner et al., 2005; Sager, 2001; van Lente et al., 2003). A striking feature is that only one project was undertaken in a low-income country (Nguyen et al., 2014). Another one has a focus on India but the project was developed in Germany (Gnatzy and Moser, 2012).

Most of the scenario projects described in the review were developed after the year 2000 and address a wide range of topics, from regional institutional perspectives (e.g. local public health departments; Venable et al., 1993) to global challenges (e.g. future infectious disease threats to Europe; Suk and Semenza, 2011). Many of the projects provide a framework for determining actions in research, as well as in public policy-making; for example, it could be the basis for discussing a national dementia plan (Bijl, 1992; Bijl and Ketting, 1991; Vollmar et al., 2014) or for developing a strategy to 'revitalize' academic medicine (Awasthi et al., 2005; Clark, 2005). None of the projects has been designated as unsuccessful by the authors, which could be either a sign of the method's strength or of publication bias. In fact, Gregório and colleagues stated: 'The use of scenario analysis in a strategic thinking process has demonstrated to be of value while planning for future resources and other policy issues' (Gregorio et al., 2014: 11). Several of these scenario projects are claimed to be helpful for strategic planning and also for enabling the incorporation of expert knowledge (the qualitative 'human factor') (Bierbooms et al., 2011; Schreuder, 1995; Venable et al., 1993). Additionally, several projects use quantitative approaches to calculate the scenarios (Buchan and Seccombe, 2012; Islei et al., 1999; Vollmar et al., 2014). The resulting scenarios are illustrated in many different ways or combinations [e.g. tables (Awasthi et al., 2005), text descriptions (Ling and Hadridge, 2000), pictures (Clark, 2005), or short stories called storylines (Vollmar et al., 2014)]. Although there is no definite response to the question of how many scenarios are optimal in the scenario planning literature (Amer et al., 2013), three to five scenarios are considered appropriate by most of the researchers (Amer et al., 2013; Pillkahn, 2007). Sometimes doubts may arise with respect to the reliability of the scenarios because the methods are not clearly described (Awasthi et al., 2005; Bezold and Peck, 2005; Bishop et al., 2007; Clark, 2005; Ling and Hadridge, 2000; Postma and Liebl, 2005). Compared with conventional methodological reporting (i.e. in clinical studies), the method in scenario projects should be described as precisely as possible due to the process-oriented character of scenario development. This includes the selection of the experts, the applied software tools, the use of additional literature sources, and also the method's use in combination with other methods, such as the Delphi technique (Bijl, 1992; Bijl and Ketting 1991; van Beeck and Mackenbach, 1997; van Beeck et al., 1989; van Lente et al., 2003).

Application of the scenario technique – two examples

Using the scenario technique in a high-income country: health care for people with dementia in 2030

Based on the typology of Kosow and Gaßner (2008; see above), this example describes a systematic-formalized scenario technique from their category 3. The aim of this scenario project was to develop various projected scenarios for the healthcare delivery system's responses to the future needs of dementia care in 2030 (Vollmar et al., 2014). To reach this goal, an exemplary six-step scenario approach was utilized (Figure 11.1).

On the basis of a literature search, an interdisciplinary core team defined the areas of influence (Step 1) and discussed potential key factors to describe future scenarios (Step 2). Based on the input of 52 experts, 25 key factors with a total of 79 future projections were entered (Step 3). A comprehensive list of key factors and projections (i.e. future directions) for each scenario is provided in Table 11.1. In the consistency matrix, the participating experts estimated 3081 consistencies for each combination of two future projections (Step 4). Mathematical analysis of a theoretical 7,140,934,453,248 constellations led to 6527 consistent projection

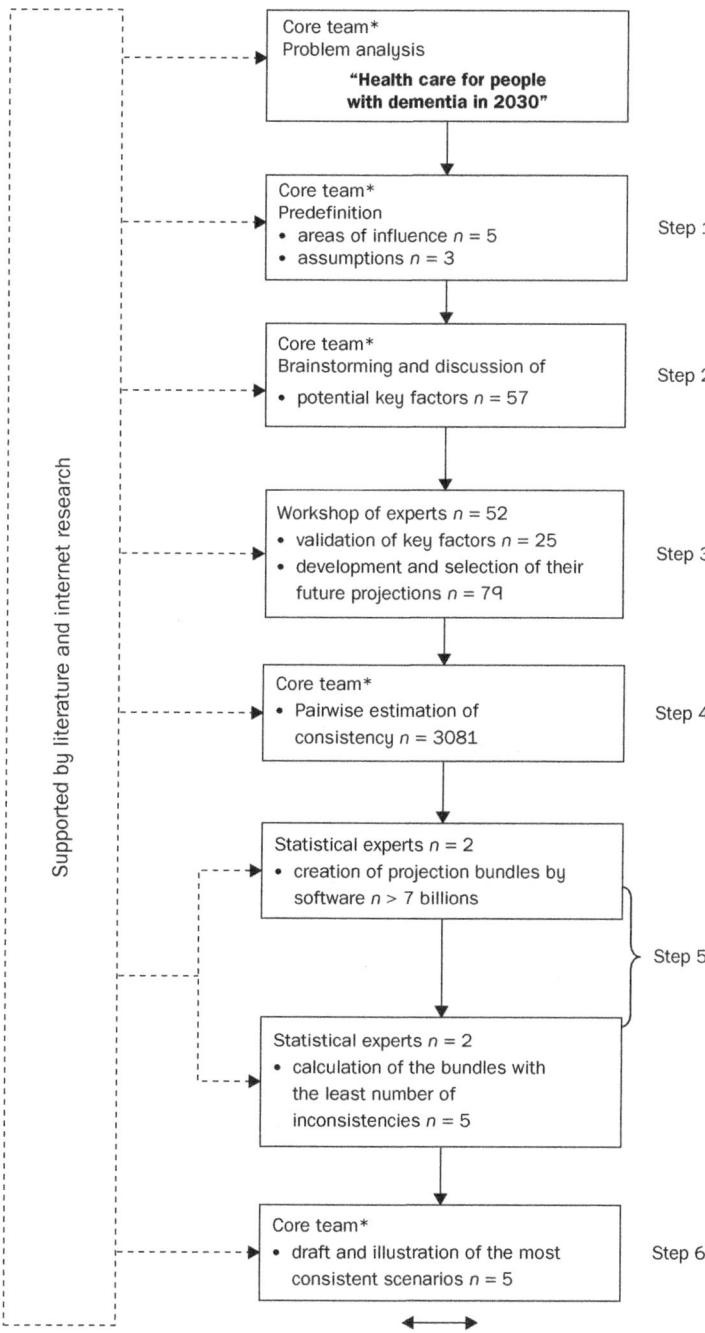

Figure 11.1 The design of one scenario project. *The core team consists of six people
Source: Vollmar et al. (2014).

bundles and simultaneously a reduction to five scenario clusters which had the highest consistency evaluations, the least number of inconsistencies, and which were coherent and plausible (Step 5).[1] From these five projection bundle clusters, five scenarios were constructed and finally described (Step 6). As Table 11.1 describes the scenarios according to their elements such as the key factors, scenario-makers very often use more holistic ways of presentation to give the scenarios life. This could be pictures, matrices, storylines or brief descriptions. In this chapter, short descriptions are provided.

Brief description of the resulting five scenarios

Scenario 1: collapse of supply structures

The economic situation in Germany is critical. The problems arising from the continuing recession dominate, forcing dementia care into the background. Dementia is no longer on the political agenda and the media pay very little attention to the subject. Only very few dementia-specific research activities are being conducted and few new findings are published. The healthcare delivery system in Germany suffers from a chronic lack of staff trained in dementia care. In addition, citizens' involvement is lacking in society. The state tries to deal with the lack of human resources by introducing statutory obligations. Each citizen is compelled to do voluntary work or pay a financial contribution. A mandatory insurance only ensures the minimum of care services, which means that admission into a nursing home is out of the question until the level of care required becomes exceedingly high. For this reason, there are hardly any alternative types of accommodation other than private homes, and persons with dementia (PwD) are cared for by relatives and friends in entirely private caring arrangements for as long as possible. Furthermore, many households employ illegal nursing staff – from countries outside Europe – because they themselves are not in a position to take on caring duties in addition to their work and other family obligations. Since caring for PwD in Germany is expensive and of poor quality, many of those concerned choose healthcare services abroad.

Scenario 2: safekeeping of people with dementia

Persons with dementia are stigmatized by society. Although health care is predominantly privately financed, the majority of PwD are cared for under 'professional' care arrangements. There is sufficient personnel available to care for dementia patients but these personnel are insufficiently qualified for the job, so that the healthcare concepts in the institutions and home care services are predominantly aimed at the safekeeping of PwD. Offers of social support and care for PwD and their relatives are inadequate and do not meet requirements. There are hardly any alternatives to living in private accommodation or in a nursing home. Good technological equipment serves mainly for the safekeeping of PwD. In spite of legal restrictions on immigration, the high costs of private care and the low quality of dementia care lead to a developing black market for foreign nursing and caring staff, with whom cost-intensive admission into a nursing home can be avoided for as long as possible. Health service/supply offers from abroad are often accepted, although these must usually be privately financed.

Table 11.1 Overview of key factors and their future projections

Key factors ('descriptors', driving forces)	Future projections (of the key factors)	Scenarios*				
		1	2	3	4	5
Dementia-specific interventions and therapies	Significant improvement	–	–	–	–	–
	Improvement	–	–	–	+	+
	No improvement	+	+	+	–	–
Dementia-specific preventative measures	Nationwide implementation of effective preventative measures	–	–	–	+	–
	Partial implementation of effective preventative measures	–	–	–	–	+
	Lack of effective preventative measures	+	+	+	–	–
Availability of assistive technologies that potentially increase the recipient's autonomy or support care	Generally available, good technological basic equipment	–	–	–	–	+
	Not generally available, good technological basic equipment	–	+	+	+	–
	Deficient technological basic equipment	+	–	–	–	–
Application of technologies	Nationwide application of technology in all areas	–	–	–	–	+
	Nationwide application, especially in a domestic environment	–	–	–	–	–
	Nationwide application, especially in a public environment	–	–	–	+	–
	Limited application in all areas	+	+	+	–	–
Risk assessment	Nationwide risk assessment	–	–	–	+	+
	Limited possibilities to assess risk	+	–	–	–	–
	No nationwide risk assessment	–	+	+	–	–
Efficiency and implementation of dementia research	Research being conducted and implemented efficiently	–	–	–	+	+
	Research being conducted efficiently but results implemented inefficiently	–	+	+	–	–
	Inefficient research	+	–	–	–	–

(Continued)

Table 11.1 continued

Key factors ('descriptors', driving forces)	Future projections (of the key factors)	Scenarios*						
Social support systems	'Mixed structure'	–	–	–	–	–	+	+
	'Go structure' (system where helpers actively go and visit people affected)	–	+	+	–	–	+	–
	'Come structure' (system where people in need of help have to actively go and utilize services)	–	–	–	–	–	–	–
	Insufficient support services	+	+	–	–	–	–	–
Care of people with dementia provided by foreign workers	Employment of foreigners is legally regulated and commonly used	–	–	–	–	–	–	+
	Employment of foreigners is legally restricted but commonly used	+	+	+	+	+	–	–
	Employment of foreign workers does not take place	–	–	–	–	–	–	–
Care of people with dementia abroad	Care abroad is legally regulated and commonly used	+	–	–	–	–	–	–
	Care abroad is not legally regulated but commonly used	+	+	+	–	–	–	–
	Care abroad is not commonly used	–	–	–	+	+	+	+
Quality and costs of care	Care is of high quality and affordable	–	–	–	–	–	+	+
	Care is of high quality and expensive	–	–	–	–	–	–	–
	Care is of affordable and of poor quality	–	–	–	–	–	–	–
	Care is expensive and of poor quality	+	+	+	+	+	–	–
Support services for people with dementia and their caregivers and/or significant other	Support services have very little or no impact	+	+	–	–	–	–	–
	Support services are effective	–	–	–	–	+	+	+
Living and care environment for people with dementia	A sufficient number of varied types of living and care environments	–	–	–	–	–	+	+
	Varied types but insufficient number of living and care environments	–	–	–	–	–	–	–
	Very limited variety of types of living and care environments	+	+	+	+	–	–	–

(Continued)

Table 11.1 continued

Key factors ('descriptors', driving forces)	Future projections (of the key factors)	Scenarios*						
Human resource situation	Appropriately trained, qualified people are available	−	−	−	−	+	+	+
	People are available but they are not sufficiently qualified	+	−	+	−	−	−	−
	Shortage in people	−	+	−	+	−	−	−
Care arrangements	Cooperative: mixed structure with professional and private care arrangements	−	−	+	+	+	+	+
	Mainly professional care arrangements	−	+	−	−	−	−	−
	Mainly private care arrangements	+	−	−	−	−	−	−
Conceptualization and implementation of care plans	Individualized care concepts are implemented in organizations in a similar manner	−	−	−	−	−	−	−
	Individualized care concepts for people with dementia are implemented in organizations differently	−	−	−	−	+	+	+
	Standardized care concepts are implemented in organizations in a similar manner	−	−	−	−	−	−	−
	Standardized care concepts for people with dementia are implemented in organizations differently	+	+	+	+	−	−	−
Percentage of people with dementia in the overall population in Germany in 2030 (prevalence)	The percentage of people with dementia in the overall population decreases or remains the same	−	−	−	−	−	−	−
	The percentage of people with dementia in the overall population increases moderately	−	−	−	−	−	−	+
	The percentage of people with dementia in the overall population increases markedly	+	+	+	+	+	+	−

(Continued)

Table 11.1 continued

Key factors ('descriptors', driving forces)	Future projections (of the key factors)	Scenarios*					
Active citizenship shown towards people with dementia	Active citizenship shown towards people with dementia is valued	−	−	−	+	+	+
	Active citizenship shown towards people with dementia is not valued, it is a state obligation for citizens to be actively involved	+	−	+	−	−	−
	The commitment citizens show towards people with dementia is not valued, there is no state involvement	−	+	−	−	−	−
Social inequality	A society with very few social differences	−	−	−	+	+	+
	Upper, middle and lower social classes with smooth transitions	−	+	−	+	+	−
	'Two-tier' society	+	−	+	+	+	−
Society's perception of dementia	Dementia as a 'normal' part of ageing	−	−	−	−	−	−
	Dementia as a 'normal' illness	−	−	−	+	+	+
	Dementia as an illness that is taboo	+	−	−	−	−	−
	Dementia as a stigmatized illness	−	+	+	−	−	−
Acceptance and motivation to utilize innovative care strategies	Utilization is normative	+	−	−	−	+	+
	Utilization is legally sanctioned	−	−	−	+	+	−
	Low utilization	−	+	+	−	−	−
Power of influence/autonomy of people affected	People affected have ways to strongly influence their situation and they take advantage of this	−	−	−	−	−	−
	People affected have in theory a variety of ways to influence their situation but only a few people take advantage of this	−	−	−	+	+	+
	People affected have hardly any ways to influence their situation	+	+	+	−	−	−

(Continued)

Table 11.1 continued

Key factors ('descriptors', driving forces)	Future projections (of the key factors)	Scenarios*				
Portion of the costs covered by private households in relation to the overall costs	Less than 50% of the costs of care are covered by private households	–	–	–	+	–
	Between 50 and 75% of the costs of care are covered by private households	–	–	–	+	+
	More than 75% of the costs of care are covered by private households	+	+	+	+	–
Structure of payers	Compulsory insurance covers a maximum of care ('Maximalversorgung': beyond standard, basic care)	–	–	+	–	–
	Compulsory insurance covers a standard, basic care; everything needed beyond has to be covered privately	–	+	+	+	+
	Compulsory insurance covers a minimum of care; everything needed beyond has to be covered privately	+	–	–	–	–
Germany's overall economic development	Economic growth improves	–	–	–	–	–
	Economic growth remains the same	–	–	–	+	+
	Recession	+	+	+	–	–
Costs of illness related to dementia	Costs of illness remain at about €10 billion	–	–	–	–	–
	Costs of illness increase up to about €15 billion	–	–	+	–	+
	Costs of illness increase to more than €15 billion	+	+	–	+	–

Note: Developed during a scenario project about health care for people with dementia in 2030 (Vollmar et al., 2014).

*Scenario 1: collapse of supply structures; Scenario 2: safekeeping of people with dementia; Scenario 3: well meant, but badly done; Scenario 4: avoiding dementia; Scenario 5: mastering dementia.

Scenario 3: well meant, but badly done

Dementia research is politically promoted and sponsored. But any new findings are inadequate and not sustainably implemented. Political programmes and efforts have no effect and the healthcare delivery system is insufficient to meet needs. Similarly, new models of dementia care are rarely implemented. Persons with dementia are stigmatized by society while citizens rarely involve themselves voluntarily on their behalf. The healthcare delivery service for PwD, which consists of a mix of private and professional caring arrangements, is supported by the state making it legally binding for citizens to work voluntarily or to make a financial contribution. There are sufficient personnel available to care for PwD but they lack dementia-specific qualifications. The individual needs and habits of those concerned are rarely taken into consideration. Many of those affected decide to employ a health service abroad because of the high costs and poor quality of the health service in their own country, where private care is also expensive.

Scenario 4: avoiding dementia

In the year 2030, dementia research receives state funding. Efficient research leads to the use of effective health services and therapies that help to improve the quality of life of those affected. Research findings are implemented sustainably and nationwide for the care of PwD through the use of political implementation programmes. All citizens are obliged by law to undergo a risk analysis and those with an increased risk of dementia must be vaccinated as a preventative measure. Thus, the relative increase of PwD in the total population can be avoided; consequently, however, citizens' right for self-determination is restricted. The healthcare delivery system has sufficient well-qualified staff in various professions to take care of PwD. Citizens' involvement in the care of PwD is highly valued in society. There are many types of accommodation, offering adequate alternatives to private homes and residential nursing homes. Technologies promoting autonomy and supporting health services in caring for PwD have also been developed but can only be found nationwide in the public sector; the expression 'barrier-free' has been extended to include PwD. Local infrastructure has been improved, which now supports the implementation of the many forms of alternative living accommodation for PwD.

Scenario 5: mastering dementia

In the year 2030, dementia has become a 'normal' disease, accepted by society. Research in all sectors of dementia is promoted. Political programmes are realized and ensure the sustainable implementation of research findings. The possibilities for risk assessment have been developed further and are applied nationwide. Effective dementia-specific preventative measures, such as voluntary vaccination, have been identified and are on offer. The social support systems for PwD and their caregivers display a mixed structure of 'outreach and walk-in model'. The various health services available for PwD are largely effective and through their individual, needs-oriented application, improve the quality of life of those concerned. Cooperative caring arrangements by private and professional persons provide health services for PwD. Citizens' involvement is highly valued in society and is integrated in health services for PwD. In addition, the influx of carers and nursing staff from abroad has been legally regulated by the government through the introduction of a Green Card. Foreign workers can be employed both in public institutions and private households, thus together with volunteers constitute sufficient multi-professional personnel. There is adequate technology available to support health services and to promote autonomy for PwD. Qualified personnel enable the use of technologies such as 'the intelligent house' (fully networked and 'anticipating' the needs of the inhabitants, e.g. automatic lights to guide PwD from bedroom to restroom) to assist the health service in the care of PwD. This also provides an assortment of living accommodation in the community for PwD so that the transition from private accommodation to a residential nursing home is a smooth process. Public sector technology provides a communal needs-oriented infrastructure for PwD, so that they can remain in their familiar surroundings for as long as possible.

Dissemination of the results

The results of this multidisciplinary scenario process (the five scenarios) provide a framework for determining actions in research as well as in society, and policy based on a list of recommendations (Vollmar et al., 2014). The recommendations were widely disseminated by internet, articles, press releases, oral presentations, and a book (Vollmar, 2014).

Using the scenario technique in a developing country: community development in Vietnam

Following the typology of Kosow and Gaßner (2008), this project used a creative-narrative scenario technique. Nguyen and colleagues 'developed scenario planning as an action-research tool in a peri-urban community' in Vietnam (Nguyen et al., 2014). Their aim was to improve a complex sanitation problem and to address emerging infectious diseases. They conducted four focus groups with eight experts from the community and developed a best- and a worst-case scenario (limited only to the best and the worst scenarios owing to time constraints in their project) (Nguyen et al., 2014). The results are shown in a matrix (Figure 11.2). During the scenario process (more precisely within the first focus group), the participants identified, discussed, and ranked seven key factors (i.e. driving forces):

1. Awareness and behavioural change of individuals (health and environment – general hygiene, keeping the house clean, washing hands with soap, disposing of garbage with health in mind).
2. Clear guidance from party and government.
3. Pollution of water resources (water in rice paddies contaminated with pesticides, household wastewater, irrigation water – Nhue River, contamination of ground water with arsenic).
4. Technological developments might be applied to address industrial waste and household waste.

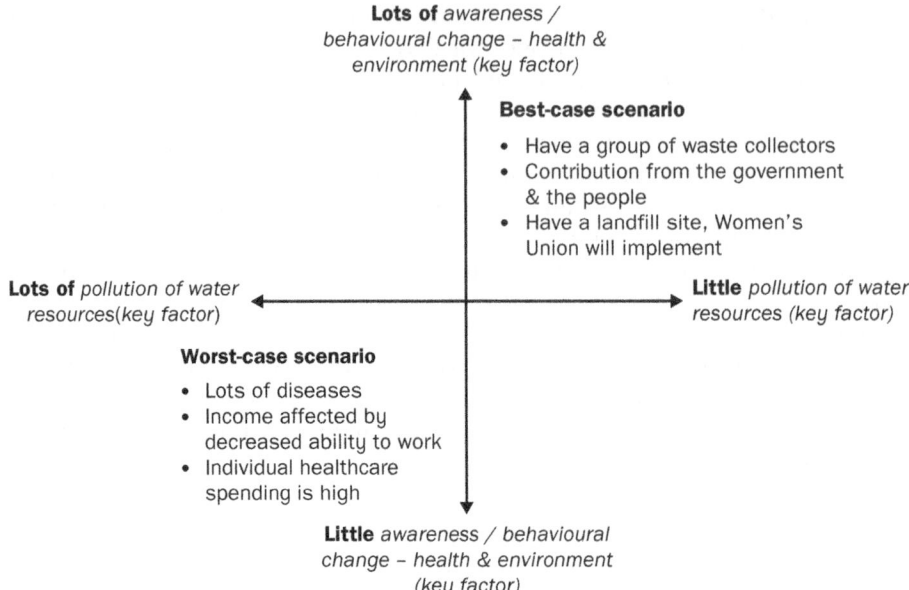

Figure 11.2 Results of one scenario project
Source: Nguyen et al. (2014).

5. Industrial development influences the number of factories.
6. Access to information – farmers lack computer skills to access information on the Internet.
7. Developments in the commune – capability to invest in solid waste management strategies.

In the second focus group meeting, participants identified the future directions (syn. projections) of the key factors. The two highest ranked key factors became the axes of the matrix (Figure 11.2).

In the third focus group meeting, participants discussed the implications of the scenarios and possible options and next steps. The main outcome of this third meeting was a specific strategy to address garbage issues in the local area. In the fourth focus group meeting, participants discussed their understanding of the process (Nguyen et al., 2014). The authors described the scenario technique as 'a powerful communication tool that can be used to convene people to discuss a shared issue' (Nguyen et al., 2014). Engagement in the scenario process led to a shift in thinking about the issue and the technique (on the part of both researchers and participants).

Conclusions

Despite its potential, published research on the scenario method is limited compared with other methods such as the Delphi technique, at least in the field of health care. This might be due to the apparent complexity of the methodological

Table 11.2 Proposed key methodological criteria to report in scenario projects

Criteria

Aim of the scenario project

- Do the words 'scenario project (planning, approach . . .)' appear in the title along with the topic of primary interest?
- Is the topic of interest clearly described?
- What are the proposed implications?
- Are the target groups and/or stakeholders specified?
- Is there a clear time horizon?

Framework of the scenario project

- Are the preconditions and presuppositions well described?
- Is the process of developing the areas of influence, key factors, and future projections adequately described?
- Who is involved (description of scenario development team and participants/experts)?
- Is the background of participants/experts clear?
- How will participants/experts be selected or excluded?

Methodological approach of the scenario project

- Is the specific scenario technique used (e.g. only narrative, consistence analysis, cross-impact analysis) well described?
- If any, is the mathematical approach well described?
- How is the mathematical approach transformed/implemented in software (if applicable)?
- Is there any combination with other methods, such as the Delphi technique?
- Is the presentation of the scenario development process adequate?
- Are the scenarios presented in a sound manner (to the specified target groups/ stakeholders)?

Impact of the scenario project

- Are there any recommendations for different target groups/stakeholders?
- What are the next steps after the scenario project?

Source: adapted from Vollmar et al. (2015).

approach. Individual project methods and activities vary widely (as shown in the two examples). Improved criteria are required for the reporting of scenario project methods. Proposed criteria are listed in Table 11.2. Only if there is transparency to reproduce the underlying evidence will the scenario method become a useful tool for future healthcare planning and strategic public health decision-making. With improved standards and greater transparency, the scenario method could become an excellent tool for scientific healthcare planning and strategic decision-making in public health.

Note

1 This was undertaken by a mathematician in this project, but there are several software packages on the market to support scenario-makers.

Recommended further reading

Scenario techniques in general

- Millennium Project: Scenarios (Glenn and The Futures Group International, 2011).
- Methods of Future and Scenario Analysis (Kosow and Gaßner, 2008).
- The current state of scenario development: an overview of techniques (Bishop et al., 2007).
- OECD: Scenario development: a typology of approaches (van Notten, 2006).

Scenario planning – How to?

- Government of the United Kingdom: Futures Toolkit: tools for strategic futures for policy-makers and analysts [https://www.gov.uk/government/publications/futures-toolkit-for-policy-makers-and-analysts] (HM Government, 2014).
- Explorer's Guide by Shell [https://www.kth.se/social/files/54e34a62f2765456cb68310e/explorers-guide.pdf] (Grundy, 2008).
- Future workshops and scenario planning (for schools and universities, only in German language) (Albers and Broux, 1999).

Examples of using the scenario technique for healthcare issues

- Public Health Agency of Canada: Children and Physical Activity Scenarios Project: evidence-based visions of the future (Butler-Jones and Trumble Waddell, 2011).
- WHO Collaborating Centre for Health Workforce Policy and Planning: A scenario-planning approach to human resources for health: the case of community pharmacists in Portugal (Gregorio et al., 2014).
- Hanoi School of Public Health (and others): Scenario planning for community development in Vietnam (Nguyen et al., 2014).
- German Centre of Neurodegenerative Disease (DZNE): Health care for people with dementia in 2030 (Vollmar, 2014; Vollmar et al., 2014).

References

Albers, O. and Broux A. (1999) *Zukunftswerkstatt und Szenariotechnik*. Weinheim: Beltz.

Amer, M., Daim, T.U. and Jetter, A. (2013) A review of scenario planning, *Futures*, 46: 23–40.

Anderson, L.A. and McConnell, S.R. (2007) The healthy brain and our aging population: translating science to public health practice, *Alzheimer's and Dementia*, 3 (2 suppl.): S1–S2.

Awasthi, S., Beardmore, J., Clark, P., Hadridge, P., Madani, H., Marusic, A. et al. (2005) Five futures for academic medicine, *PLoS Medicine*, 2 (7): e207 [doi: 10.1371/journal.pmed.0020207].

Baker, J., Lovell, K. and Harris, N. (2006) How expert are the experts? An exploration of the concept of expert within Delphi panel techniques, *Nurse Researcher*, 14 (1): 59–70.

Becker, H.A. (1988) Simulating the aging of the Netherlands, *Simulation and Games*, 19 (2): 186–209.

Bezold, C. (1992) Five futures, *The Healthcare Forum Jorunal*, 35 (3): 28–42.

Bezold, C. (2005) The future of patient-centered care: scenarios, visions, and audacious goals, *Journal of Alternative and Complementary Medicine*, 11 (suppl. 1): S77–S84.

Bezold, C. and Peck, J. (2005) Drug regulation 2056, *Food and Drug Law Journal*, 60 (2): 127–36.

Bierbooms, J.J., Bongers, I.M. and van Oers, H.A. (2011) A scenario analysis of the future residential requirements for people with mental health problems in Eindhoven, *BMC Medical Informatics and Decision Making*, 11: 1 [doi: 10.1186/1472-6947-11-1].

Bijl, R. (1992) Delphi in a future scenario study on mental health and mental health care, *Futures*, 24 (3): 232–50.

Bijl, R.V. and Ketting, E. (1991) [Future scenario's Dementia 1990–2010: various major points from the scenario report 'Care for mental health in the future'], *Tijdschrift voor Gerontologie en Geriatrie*, 22 (3): 110–14; discussion 114–16.

Bishop, P., Hines, A. and Collins, T. (2007) The current state of scenario development: an overview of techniques, *Foresight*, 9 (1): 5–25.

Buchan, J. and Seccombe, I. (2012) Using scenarios to assess the future supply of NHS nursing staff in England, *Human Resources for Health*, 10 (1): 16 [doi: 10.1186/1478-4491-10-16].

Butler-Jones, D. and Trumble Waddell, J. (2011) *Children and Physical Activity Scenarios Project: Evidence-based visions of the future*. Ottawa: Public Health Agency of Canada.

Clark, J. (2005) Five futures for academic medicine: the ICRAM scenarios, *British Medical Journal*, 331 (7508): 101–4.

Diamond, I.R., Grant, R.C. and Feldman, B.M. (2014) Defining consensus: a systematic review recommends methodologic criteria for reporting of Delphi studies, *Journal of Clinical Epidemiology*, 67 (4): 401–9.

Eberl, I. and Schnepp, W. (2006) [Family health nursing in Germany: a process of consensus finding as basis], *Pflege*, 19 (4): 234–43.

Enzmann, D.R., Beauchamp, N.J., Jr. and Norbash, A. (2011) Scenario planning, *Journal of the American College of Radiology*, 8 (3): 175–9.

Gerhold, L., Holtmannspötter, D., Neuhaus, C., Schüll, E., Schulz-Montag, B., Steinmüller, K. et al. (eds.) (2015) *Standards und Gütekriterien der Zukunftsforschung. Ein Handbuch für Wissenschaft und Praxis* (Zukunft und Forschung, 4). Wiesbaden: Springer.

Glenn, J.C. and The Futures Group International (2011) Scenarios, in J.C. Glenn and T.J. Gordon (eds.) *Futures Research Methodology Version 3.0*. The Millennium Project.

Gnatzy, T. and Moser, R. (2012) Scenario development for an evolving health insurance industry in rural India: INPUT for business model innovation, *Technological Forecasting and Social Change*, 79 (4): 688–99.

Gregorio, J., Cavaco, A. and Velez Lapao, L. (2014) A scenario-planning approach to human resources for health: the case of community pharmacists in Portugal, *Human Resources for Health*, 12: 58 [doi: 10.1186/1478-4491-12-58].

Grundy, P. (2008) *Scenarios: An explorer's guide*. The Hague: Shell International BV.

HM Government (2014) *Futures Toolkit: Tools for strategic futures for policy-makers and analysts*. London: Cabinet Office and Government Office for Science [available at: https://www.gov.uk/government/publications/futures-toolkit-for-policy-makers-and-analysts; accessed 14 July 2017].

Islei, G., Lockett, G. and Naudé, P. (1999) Judgemental modelling as an aid to scenario planning and analysis, *Omega: The International Journal of Management Science*, 27 (1): 61–73.

Kahn, H. and Wiener, A.J. (1967) *The Year 2000: A framework for speculation on the next thirty-three years*. New York: Macmillan.

Karger, C.R. (2013) Citizen scenarios for the future of personalized medicine: a participatory scenario process in Germany, *International Journal of Interdisciplinary Social and Community Studies*, 7 (2): 1–16.

Kosow, H. and Gaßner, R. (2008) *Methods of Future and Scenario Analysis: Overview, assessment, and selection criteria*. Bonn: Deutsches Institut für Entwicklungspolitik/German Development Institute.

Landeta, J. (2006) Current validity of the Delphi method in social sciences, *Technological Forecasting and Social Change*, 73 (5): 467–82.

Leufkens, H., Hekster, Y. and Hudson, S. (1997) Scenario analysis of the future of clinical pharmacy, *Pharmacy World and Science*, 19 (4): 182–5.

Lexa, F.J. and Chan, S. (2010) Scenario analysis and strategic planning: practical applications for radiology practices, *Journal of the American College of Radiology*, 7 (5): 369–73.

Ling, T. and Hadridge, P. (2000) The Madingley scenarios for the future context of healthcare: understanding today by imagining tomorrow, *British Homeopathic Journal*, 89 (suppl. 1): S3–S7; discussion S8–S9.

Ma, S. and Seid, M. (2006) Using foresight methods to anticipate future threats: the case of disease management, *Health Care Management Review*, 31 (4): 270–9.

Meristo, T., Tuohimaa, H., Leppimaki, S. and Laitinen, J. (2009) Alternative futures of proactive tools for a citizen's own wellbeing, *Studies in Health Technology and Informatics*, 146: 144–8.

Moore, S., Mawji, A., Shiell, A. and Noseworthy, T. (2007) Public health preparedness: a systems-level approach, *Journal of Epidemiology and Community Health*, 61 (4): 282–6.

Neiner, J.A., Howze, E.H. and Greaney, M.L. (2004) Using scenario planning in public health: anticipating alternative futures, *Health Promotion Practice*, 5 (1): 69–79.

Nguyen, V., Nguyen-Viet, H., Pham-Duc, P. and Wiese, M. (2014) Scenario planning for community development in Vietnam: a new tool for integrated health approaches?, *Global Health Action*, 7: 24482 [doi: 10.3402/gha.v7.24482].

Nielsen, G.A. (1996) Preparing for change: strategic foresight scenarios, *Radiology Management*, March/April: 43–7.

Niewoehner, J., Wiedemann, P., Karger, C., Schicktanz, S. and Tannert, C. (2005) Participatory prognostics in Germany – developing citizen scenarios for the relationship between biomedicine and the economy in 2014, *Technological Forecasting and Social Change*, 72 (2): 195–211.

O'Cathain, A., Murphy, E. and Nicholl, J. (2008) The quality of mixed methods studies in health services research, *Journal of Health Services Research and Policy*, 13 (2): 92–8.

Pillkahn, U. (2007) *Trends und Szenarien als Werkzeuge zur Strategieentwicklung. Wie Sie die unternehmerische und gesellschaftliche Zukunft planen und gestalten.* Erlangen: Publicis Kommunikations Agentur GmbH, GWA.

Postma, T.J.B.M. and Liebl, F. (2005) How to improve scenario analysis as a strategic management tool?, *Technological Forecasting and Social Change*, 72 (2): 161–73.

Retel, V.P., Joore, M.A., Linn, S.C., Rutgers, E.J. and van Harten, W.H. (2012) Scenario drafting to anticipate future developments in technology assessment, *BMC Research Notes*, 5: 442 [doi: 10.1186/1756-0500-5-442].

Rhea, M. and Bettles, C. (2012) Four futures for dietetics workforce supply and demand: 2012–2022 scenarios, *Journal of the Academy of Nutrition and Dietetics*, 112 (3 suppl.): S25–S34.

Rydstrom, C. and Tornberg, S. (2006) Cervical cancer incidence and mortality in the best and worst of worlds, *Scandinavian Journal of Public Health*, 34 (3): 295–303.

Sager, B. (2001) Scenarios on the future of biotechnology, *Technological Forecasting and Social Change*, 68: 109–29.

Schaapveld, K. and Cleton, F.J. (1989) Cancer in The Netherlands: from scenarios to health policy, *European Journal of Cancer and Clinical Oncology*, 25 (4): 767–71.

Schreuder, R.F. (1995) Health scenarios and policy making: lessons from the Netherlands, *Futures*, 27 (9/10): 953–8.

Schwartz, P. (1996) *The Art of the Long View: Paths to strategic insight for yourself and your company.* New York: Doubleday.

Suk, J.E. and Semenza, J.C. (2011) Future infectious disease threats to Europe, *Americal Journal of Public Health*, 101 (11): 2068–79.

van Beeck, E.F. and Mackenbach, J.P. (1997) Future health scenarios as a tool in the surveillance of unintentional injuries, *Health Policy*, 40 (1): 13–28.

van Beeck, E.F., Mackenbach, J.P., van Oortmarssen, G.J., Barendregt, J.J.M., Habbema, J.D.F. and van der Maas, P.J. (1989) Scenarios for the future development of accident mortality in The Netherlands, *Health Policy*, 11 (1): 1–17.

van der Heijden, K., Bradfield, R., Burt, G., Cairns, G. and Wright, G. (2002) *The Sixth Sense: Accelerating organizational learning with scenarios*. New York: Wiley.

van Lente, H., Willemse, J., Vorstman, C. and Modder, J.F. (2003) Scenario planning as policy instrument: four scenarios for biotechnology in Europe, *Innovation: Management, Policy and Practice*, 5: 4–14.

van Notten, P.W.F. (2006) Scenario development: a typology of approaches, in OECD (ed.) *Think Scenarios, Rethink Education*. Paris: OECD Publishing.

Varum, C.A. and Melo, C. (2010) Directions in scenario planning literature – a review of the past decades, *Futures*, 42 (4): 355–69.

Venable, J.M., Ma, Q.L., Ginter, P.M. and Duncan, W.J. (1993) The use of scenario analysis in local public health departments: alternative futures for strategic planning, *Public Health Reports*, 108 (6): 701–10.

Vollmar, H.C. (ed.) (2014) *Leben mit Demenz im Jahr 2030: Ein interdisziplinäres Szenario-Projekt zur Zukunftsgestaltung*. Versorgungsstrategien für Menschen mit Demenz. Weinheim: Beltz Juventa.

Vollmar, H.C., Goluchowicz, K., Beckert, B. (2014) Health care for people with dementia in 2030 – results of a multidisciplinary scenario process, *Health Policy*, 114 (2/3): 254–62.

Vollmar, H.C., Ostermann, T. and Redaelli, M. (2015) Using the scenario method in the context of health and health care – a scoping review, *BMC Medical Research Methodology*, 15: 89 [doi: 10.1186/s12874-015-0083-1].

Wiek, A., Gasser, L. and Siegrist, M. (2009) Systemic scenarios of nanotechnology: sustainable governance of emerging technologies, *Futures*, 41 (5): 284–300.

Wilkinson, A., Kupers, R. and Mangalagiu, D. (2013) How plausibility-based scenario practices are grappling with complexity to appreciate and address 21st century challenges, *Technological Forecasting and Social Change*, 80 (4): 699–710.

Zentner, R.D. (1991) Scenarios: a planning tool for health care organizations, *Hospital and Health Services Administration*, 36 (2): 211–22.

12 Outcome mapping: a tool for planning, monitoring, and evaluating complex interventions in systems

Jenna M. Evans

Introduction

What is outcome mapping?

Outcome mapping is a method used to plan, monitor, and evaluate complex interventions. It consists of a set of tools and guidelines that bring various partners together to identify their desired results and the behaviours and activities needed to achieve those results. By closely monitoring changes in the identified behaviours and activities, the contributions of the intervention towards achieving the desired results can be assessed.

Outcome mapping is well- suited to answer the following questions (Williams and Hummelbrunner, 2011):

- How does our intervention contribute to an ultimate goal?
- Whose behaviours can we influence to support that contribution?
- What is a realistic strategy to achieve that behaviour change?
- How can we assess behaviour change?

Distinguishing features of outcome mapping

Outcome mapping has three features that distinguish it from conventional evaluation methods and tools. First, it recognizes that multiple, often non-linear causes lead to change, and that interventions contribute to change, but are not necessarily responsible for it. This is in stark contrast to tools such as the logframe or means-end diagram, which conceptualize change as a linear, one-way cause-and-effect process.

Second, outcome mapping incorporates the roles and perspectives of multiple stakeholders involved in or affected by the intervention. It relies on dialogue among stakeholders and consensus-building to create a shared vision and shared understanding of how their individual and collective activities contribute to intended outcomes. Few evaluation frameworks and tools explicitly involve stakeholder engagement.

Finally, outcome mapping assesses an intervention in terms of its influence on the people, groups, and organizations involved in or affected by the intervention.

The outcomes of interest are changes in the behaviours, relationships or activities of these stakeholders. Conventional evaluation methods, on the other hand, typically measure deliverables and impact on primary beneficiaries using pre-defined quantitative performance indicators. This results-based approach can hinder flexibility, learning, and innovation, as it is frequently unclear how and why impact occurs.

Value of outcome mapping to health systems research and practice

Ageing populations, the growing burden of chronic disease, new technologies and treatments, and financial constraints have led health systems around the world to seek fundamental changes in how health care is organized, financed, and delivered (Shortell et al., 2000). Many of these reform efforts are aimed at promoting care that is better integrated across diverse healthcare sectors, organizations, and professional (Wodchis et al., 2015). Healthcare managers and clinicians are thus under increasing pressure to work together collaboratively and to consider the performance of the healthcare system, not just their respective teams and organizations. In the context of these transformative changes, outcome mapping is a valuable tool for researchers and leaders.

Outcome mapping facilitates the planning and implementation of complex healthcare interventions by clarifying who the key stakeholders are, what their relationships are, and how their work contributes to desired outcomes. Given the multitude of diverse organizations and professionals involved in healthcare delivery, identifying an intervention's stakeholders and sphere of influence is important for determining where attention and resources should be focused to support behaviour change and outcome achievement.

The process of outcome mapping engages stakeholders, enabling them to forge or strengthen relationships and build a shared understanding of what their respective roles are and how they can collectively achieve their goals. Healthcare leaders and clinicians rarely have the opportunity to engage in the type of inter-professional and inter-organizational dialogue, reflection, and problem-solving that outcome mapping entails. Furthermore, by involving stakeholders in the planning and monitoring process and emphasizing reflection on relationships and responsibilities, outcome mapping fosters ownership and accountability for the intervention and its outcomes.

Outcome mapping supports higher-level decision-making and learning. It serves as a decision-making aid by illuminating potential pathways for change and facilitating the examination of interdependencies and trade-offs between different options (Ehrlich et al., 2015; Garfinkel et al., 2006; Tsasis et al., 2013). The cyclical and iterative nature of outcome mapping fosters learning about the stakeholders, contexts, and challenges involved in complex interventions.

Finally, outcome mapping complements dominant approaches to performance management and measurement in health care, such as the balanced scorecard (BSC) and structure-process-outcomes (SPO) Model. The BSC focuses attention on balancing financial and non-financial performance indicators across four quadrants: learning and growth, internal processes, patient outcomes, and financial outcomes (Inamdar et al., 2002). The SPO model, used widely to assess quality of care, organizes performance indicators into three categories: structures that describe the context in which care is delivered; processes that describe how care is delivered; and outcomes, which refer to the effects of health care on the health status of patients and populations (Donabedian, 1978). The BSC and SPO model enable managers to assess performance from multiple dimensions to support ongoing evaluation and improvement. However, these measurement systems have difficulty explaining the interactions and interdependencies among the various organizations and stakeholders that shape healthcare delivery and health outcomes. They also encourage a narrow focus on activities and outcomes that are well defined and quantifiable, even though some important activities and outcomes may not be easily measured. Outcome mapping addresses these limitations and can be used alongside frameworks such as the BSC and SPO model.

Although originally developed for use in the international development sector, outcome mapping is increasingly used to enhance health systems research and practice. In South Africa, senior managers at a primary healthcare clinic used outcome mapping to create and assess clinical care teams aimed at improving collaboration between doctors and nurse practitioners (Mash et al., 2008). In Australia and Canada, it has been used to build consensus among diverse partners with the aim of integrating care across professional, organizational, and sectoral boundaries (Ehrlich et al., 2015; Tsasis et al., 2013). Outcome mapping has also been used in the United States as a decision support tool to identify and select policy options and research opportunities in the field of perinatal health (Garfinkel et al., 2006) and to support performance management and accountability (Persaud and Nestman, 2006).

The remainder of this chapter is organized into two sections. In the first, the outcome mapping methodology is explained. Then, in the second, two case studies are provided that describe applications of outcome mapping in Canadian and South African healthcare contexts.

Methodology

Steps of outcome mapping

Outcome mapping consists of twelve steps divided into three stages: (1) intentional design, (2) outcome and performance monitoring, and (3) evaluation planning. The first stage, *intentional design*, helps establish consensus on the macro-level changes the intervention intends to achieve. The desired changes are conceptualized and discussed at three interrelated levels: the implementation team, the boundary partners, and the environment.

This stage involves the following seven steps:

1. Develop a clear, concise, and inspiring vision statement that describes why the intervention exists.
2. Develop a mission statement that describes how the intervention will contribute to the vision. The mission statement outlines the areas in which the intervention will focus.
3. Identify boundary partners. Boundary partners are those individuals, groups or organizations with whom the intervention interacts directly and who can contribute to the vision.

4. Identify the outcomes boundary partners need to achieve in order to realize the vision. Outcomes are defined as changes in the behaviours, relationships or activities of boundary partners.
5. Develop progress markers for each outcome. Progress markers are statements describing the gradual progression of changed behaviour of boundary partners leading to the desired outcomes. Progress markers for each outcome reflect what one would *expect to see* the boundary partner doing as an early response to the intervention, what one would *like to see* them doing, and what one would *love to see* them doing if the intervention were having a profound influence.
6. Identify strategies the implementation team will use to contribute to and support change among boundary partners. These strategies may be aimed directly at the boundary partners and/or at the environment in which the intervention and boundary partners operate.
7. Identify organizational practices that explain how the intervention and implementation team are going to operate in order to fulfil their mission.

Although not a formal part of the outcome mapping methodology, some applications of outcome mapping result in the production of an outcome map (e.g. Garfinkel et al., 2006; Tsasis et al., 2013). The outcome map visually depicts how particular strategies and organizational practices (shown as rectangles in the map) contribute to improvements in progress markers and outcomes (shown as circles) (Figure 12.1). Relationships between elements in the outcome map are illustrated using arrows. The outcome map aims to reflect reality to the greatest extent possible, so change pathways are not discrete, but rather interconnected networks of action.

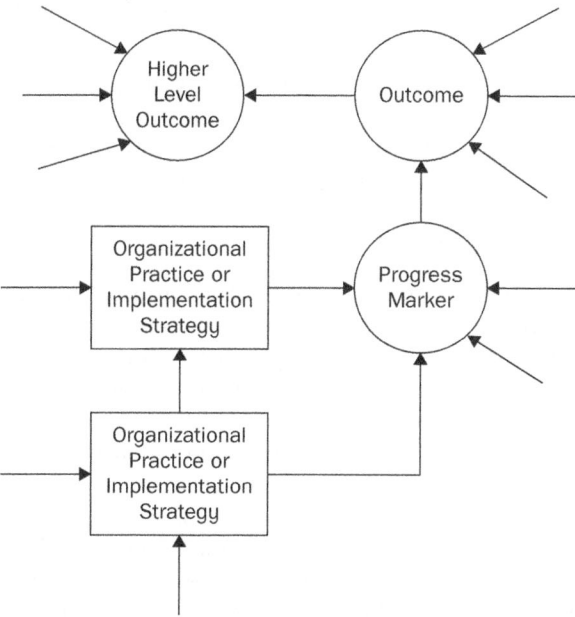

Figure 12.1 Sample structure of an outcome map

The second stage of outcome mapping, *outcome and performance monitoring*, involves the use of data collection sheets to track outcomes and change strategies and organizational practices. This stage involves the following four steps:

8. Identify priority areas to monitor on an ongoing basis and decide whether to utilize a 'light' monitoring system in which teams meet regularly to review progress or a 'heavy' monitoring system in which data are documented more extensively.

9. Create an outcomes data collection sheet to gather information about the progress markers over time. The template should include each progress marker, a description of the change exhibited, and who among the boundary partners exhibited the change. Information explaining the reasons and circumstances for the change, evidence of the change, unanticipated change, and lessons for the program are also recorded in order to keep a running track of the context for future analysis or evaluation.

10. Create a data collection sheet to record information on the strategies being used to encourage change in the boundary partners. The template should include the strategy being monitored, a description of related activities (what was done, with whom, and when?), a description of effectiveness (how did those activities influence change in one or more boundary partners?), required follow-up or additional changes, and lessons learned.

11. Create a performance data collection sheet to record data on how the intervention is operating to fulfil its mission. This self-assessment can include quantitative, qualitative or mixed data on the organizational practices used by the intervention.

The third stage of outcome mapping, *evaluation planning*, involves identifying specific aspects of the intervention to be studied and assessed in depth. This stage consists of the following step:

12. Develop an evaluation plan that describes the main elements of the evaluation to be conducted. The plan should outline priority evaluation topics and questions, a utilization strategy for the evaluation findings, the person(s) responsible for conducting the evaluation, and the projected cost.

Stakeholder consultation

Outcome mapping encourages a participatory approach in the completion of its twelve steps. Stakeholders involved in or affected by the intervention should have the opportunity to share their experiences and views to inform the identification of key boundary partners, outcomes, progress markers, change strategies, and organizational practices. These stakeholders should also provide input into decisions made as part of the monitoring and evaluation stages.

Various formal methods for stakeholder consultation and consensus-building exist, which can ensure a systematic, rigorous, and equitable process. The three most common methods are the nominal group technique, the Delphi method, and the consensus development conference (Jones and Hunter, 1995). The nominal group technique uses a highly structured meeting to gather information from relevant stakeholders or experts about a given issue. It consists of two rounds in which panelists rate, discuss, and then re-rate a series of items or questions. The Delphi method also solicits the opinions of stakeholders or experts, but does so through the use of a series of questionnaires combined with the provision of feedback. Typically, three data collection and feedback 'rounds' are undertaken to obtain a reliable group opinion, although in some cases the process continues until group consensus is reached. The Delphi method does not require that participants meet in person. Consensus development conferences involve face-to-face discussion and debate among stakeholders guided by a facilitator (Halcomb et al., 2008). These conferences are usually less formal than the nominal group technique or Delphi method with respect to the criteria for making decisions, definition of consensus, and voting or polling of participants (McGlynn et al., 1990).

The International Development Research Centre (IDRC) also provides a detailed guide for conducting outcome mapping workshops, complete with a description of workshop materials and tips for facilitation (Earl et al., 2001).

Use of published evidence

To complement stakeholder input, published evidence can be used to identify and support linkages between particular behaviours or activities and desired outcomes as well as to identify effective change strategies and organizational practices. A synthesis of relevant evidence can provide a starting point for stakeholder discussions about what is known and unknown about the intervention (or similar interventions) and its successful implementation. Linking change strategies and organizational practices to supporting evidence enables participants to understand the basis for particular decisions and whether decisions were based on research, expert opinion, and/or stakeholder consensus (McGlynn et al., 1990).

Practical considerations

Outcome mapping is best used from the planning stage of an intervention, but can also be used as an assessment tool (for external or self-assessment) during or at the end of an intervention. It is important to apply outcome mapping to an intervention or programme that is specific and has a well-defined scope. If too broad or too general, it is difficult to identify who will change and how they will change. Outcome mapping also requires an ongoing investment of time, funds, and human resources. Readiness for and commitment to outcome mapping should be carefully considered prior to implementation. Finally, the outcome mapping methodology has been adapted in various ways to meet the needs or preferences of its users. Although intended for use as a holistic evaluation system, the steps and tools of outcome mapping may also be applied separately with positive impact.

Applications of outcome mapping

South Africa

In 2004, the Worchester Community Health Centre (CHC) in the Cape Winelands District of South Africa attempted to create practice teams consisting of doctors and nurses with the aim of improving the delivery of primary care. The experiment failed, leading to a nine-month cooperative inquiry (Reason, 1988) to explore barriers to change and to reintroduce practice teams (Mash et al., 2007, 2008). The cooperative inquiry group consisted of provincial and municipal facility managers, the senior family physician, a principal medical officer, the primary care nurse trainer, and an administrative clerk. The group was co-facilitated by two external researchers.

The cooperative inquiry group (CIG) used outcome mapping to guide the re-implementation of practice teams. They developed a vision for the CHC: 'Our community health centre will be the prime example of quality, family-oriented primary health care and a place of hope for all who are sick.' Their mission involved delivering care that is comprehensive and patient-centred through trusting, collaborative, and continuous relationships among care providers and between providers and patients. Five boundary partner groups were identified: (1) doctors, (2) nurse practitioners, (3) senior managers, (4) administrative and support staff, and (5) patients. Examples of outcome challenges, progress markers, and strategies are provided in Table 12.1.

Four cycles of planning, action, observation, and reflection were conducted using the plan generated by outcome mapping. At the end, the CIG scored the extent to which each strategy had been successfully completed and the extent to which each boundary partner's progress markers had been achieved as high (2), medium (1) or low (0). A final score was calculated for each strategy and

Table 12.1 Examples from the Worchester Community Health Centre's outcome mapping process aimed at developing clinical practice teams, South Africa

Outcome mapping step	Example
Outcome challenges: The CIG described the change they would like to see in each boundary partner	*For doctors*: Doctors are approachable, communicate effectively with colleagues both in and outside of the CHC, act professionally, and are life-long learners. They work with the nurse practitioner as an equal partner in the provision of primary health care and endeavour to work with them in practice teams
Progress markers: The CIG defined the steps involved in achieving the outcome challenges	*For doctors*: Expect to see: Doctors give feedback to nurse practitioners on their referrals Like to see: Doctors are keen to teach and mentor the nurse practitioners on specific patients Love to see: Members of practice teams share workload and no one is overworked or over-pressured.
Strategies: The CIG defined the activities that each boundary partner would engage in to achieve the progress markers	*For doctors and nurse practitioners*: Conduct meetings with nurses and doctors together to discuss important issues for the functioning of the team, such as referral of patients and how nurses can consult doctors for support and advice

for each boundary partner expressed as a percentage of the total possible score if all were achieved. The two new practice teams and the change process were evaluated based on the results of focus groups and interviews with doctors and nurses focusing on their experience of working in a team as well as a questionnaire that assessed the quality of teamwork.

All boundary partners, except patients, achieved more than 60% of the total possible score on their progress markers. Senior managers, nurse practitioners, and administrative and support staff achieved more than 70% of their progress markers, while doctors achieved just over 60% and patients just over 50%. The CIG developed 40 specific strategies to support the implementation and success of the practice teams. Out of the 40 strategies, 27 (67.5%) were rated as 'high' in terms of successful engagement and completion, 9 (22.5%) as 'medium', and 4 (10%) as 'low'. The results of the teamwork questionnaire suggest positive team functionality, with both teams giving themselves a score of 4 or 5 (1 being poor and 5 being good) on all dimensions, including team goals, my role, procedures, decisions, managing conflict, availability, and mutual support.

Overall, the combined use of cooperative inquiry and outcome mapping helped to support organizational change in this complex, low-resource setting. For more information on this case, consult Mash et al. (2007, 2008).

Canada

In 2006, fourteen geographically defined Local Health Integration Networks (LHINs) were created in the Canadian province of Ontario. The LHINs plan, fund, and manage services within their regions using integrated health service plans developed in collaboration with local healthcare providers and community members. In 2008, the Central LHIN embarked on an outcome mapping project with the aims of creating stakeholder alignment around a shared vision and producing a comprehensive, clear, and actionable road map to guide decisions and actions (Tsasis et al., 2013).

A steering committee was created consisting of Central LHIN representatives, including a senior director, an epidemiologist, and a board member, as well as senior directors from a local hospital and a community agency and three researchers. A consensus development conference (CDC) method was used to elicit the experiences and perspectives of 45 healthcare professionals from across the healthcare sector. A CDC involves face-to-face discussion and debate between stakeholders, and is a research tool used to enhance decision-making, identify strategic directions, and advance stakeholder ownership and engagement (Halcomb et al., 2008). Participants were selected from organizations within the Central LHIN. These professionals spanned managerial (n = 18), clinical (n = 15), administrative (n = 9), and research (n = 3) roles, and represented the full continuum of care, including the LHIN, acute care, rehabilitation and complex continuing care, primary care, home care, and community support agencies.

Two independent consultants were hired to facilitate the CDCs. They asked questions focused on health system goals, indicators of success, and barriers and enablers to goal attainment. Responses were recorded on a flip chart and participants were given the opportunity to clarify, dispute, and discuss any points on the list. To ensure that responses were accurately captured, the conversations were tape-recorded and transcribed verbatim. Through analysis of the transcripts, the themes generated at the CDC were further refined and categorized into outcomes and activities for inclusion in the map. These data were used in the development of the first draft of the outcome map.

The first draft of the map was then presented to the participants at the second CDC where they were asked to comment on anything wrong or missing and what should be modified or added and why. Dialogue was focused on clarifying high-level goals and objectives, adding missing outcomes, perfecting the language used to describe outcomes, and correcting the logic of the change pathways. All comments were tape-recorded and transcribed verbatim for review in preparation of the next phase of the outcome map.

Following this CDC, all participants, including those who could not attend, were sent a copy of the map to allow for further reflection and feedback. It took an additional iteration to finalize the map in dialogue with participants and the steering committee. The outcome map represented a graphic illustration of desired outcomes and supporting actions that contribute to the functioning of the healthcare system. An excerpt from the outcome map, focused on delivering integrated care, is provided in Figure 12.2.

In the closing discussion of the final CDC, participants expressed that the outcome map highlights the complex inter-linkages present in the system, while the consultation process revealed pools of knowledge – groups and organizations – they did not know existed. Outcome mapping thus helped strengthen stakeholder awareness and alignment within a complex system, and build new relationships and commitments to collaborative work. For more information on this case, consult Tsasis et al. (2013) and Central LHIN (2009).

Commentary

The South African and Canadian case examples described above involved different interpretations and approaches to outcome mapping. In South Africa, Bob Mash and his team followed the outcome mapping methodology closely with an emphasis on measuring progress markers linked to specific behavioural outcomes exhibited by each boundary partner. This approach resulted in the development of a concrete plan of action for the implementation team as well as each boundary partner, and a means for tracking changes in behaviour. However, it is not clear whether data collection sheets, which are considered a staple of the outcome mapping methodology, were used over time to track outcomes and change strategies and organizational practices.

In Canada, Peter Tsasis and his team focused their efforts on the production of an outcome map to visually depict desired outcomes and contributing actions. The outcome map helped stakeholders achieve a shared vision and a common understanding of how the vision will be realized. However, the map does not explicitly link actions and outcomes to specific stakeholder groups (i.e. boundary partners). This oversight may make it difficult to use the outcome map in practice, as it is unclear *who* is responsible for particular actions and outcomes.

SYSTEM INTEGRATION

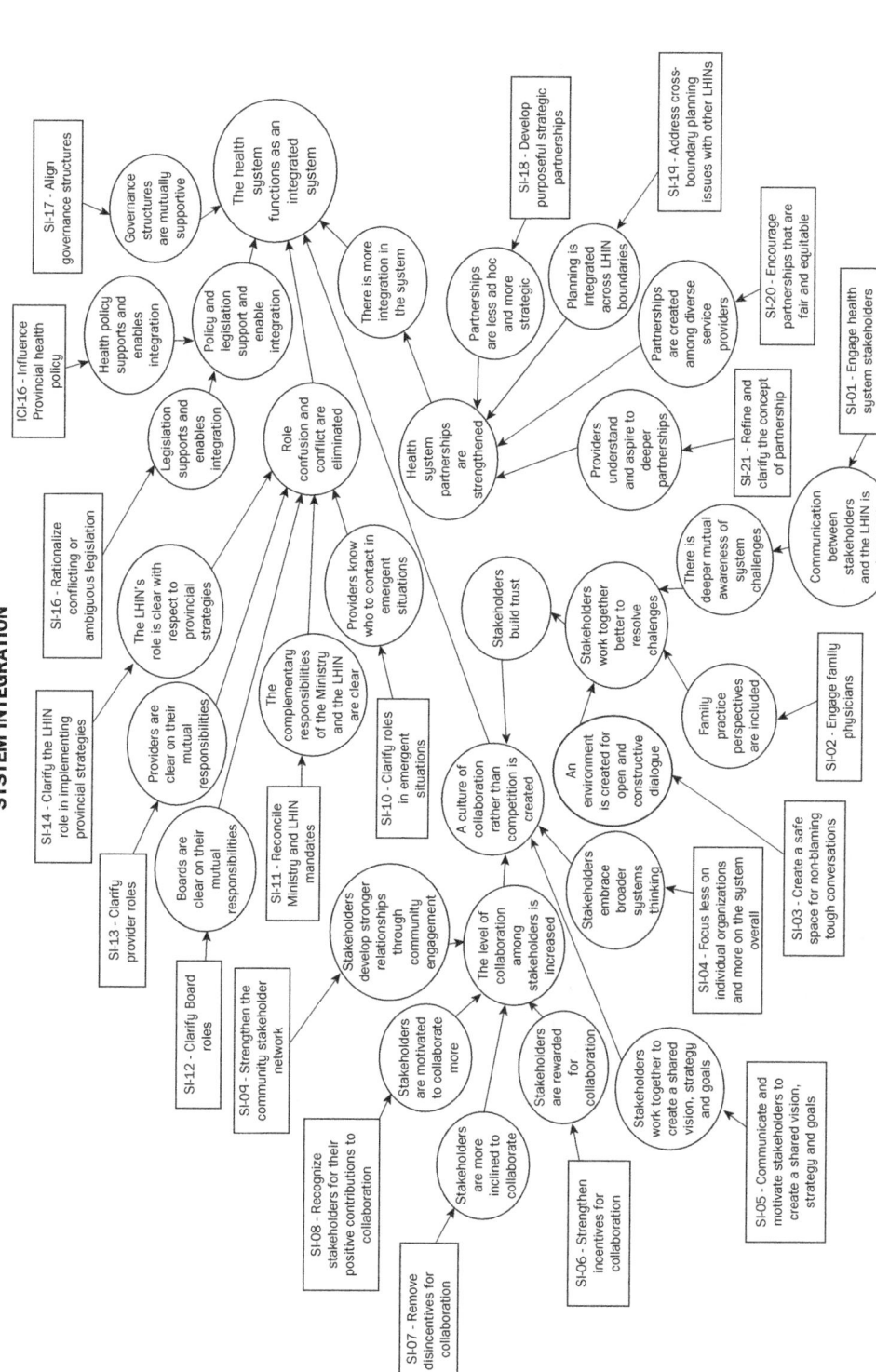

Figure 12.2 System integration outcome map for the Central Local Health Integration Network, Ontario, Canada

Source: Tsasis et al. (2013).

These two case examples demonstrate diverse approaches for applying outcome mapping in health systems. The examples highlight the importance of using outcome mapping as a holistic evaluation system. Although various steps and tools of outcome mapping may be applied separately with positive impact, the consequences of doing so must be carefully considered based on programme aims and the intended use of the data and information generated by outcome mapping.

Summary

Outcome mapping helps to clarify what an intervention wants to accomplish, with whom, and how. It also offers a method for monitoring changes in the behaviours of key stakeholders and changes in the intervention itself. As a whole, outcome mapping provides a comprehensive system for thinking holistically and strategically about how it intends to achieve results. By actively engaging key stakeholders in the planning, monitoring, and evaluation process, outcome mapping facilitates learning and reflection as well as ownership of the intervention. Outcome mapping encourages its users to think of the intervention, its context, and its stakeholders as dynamic systems whose goals, methods, and relationships need to be reconsidered and adjusted regularly to support ongoing improvement.

References

Central Local Health Integration Network (LHIN) (2009) *Central LHIN Health Integration Outcome Map* [available at: http://www.centrallhin.on.ca/Search.aspx?search=Central%20 LHIN%20Health%20Integration%20Outcome%20Map; accessed 14 July 2017].

Donabedian, A. (1978) The quality of medical care, *Science*, 200 (4344): 856–64.

Earl, S., Carden, F. and Smutylo, T. (2001) *Outcome Mapping: Building learning and reflection into development programs*. Ottawa: International Development Research Centre.

Ehrlich, C., Kendall, E., Frey, N., Dentoin, M. and Kisely, S. (2015) Consensus building to improve the physical health of people with severe mental illness: a qualitative outcome mapping study, *BMC Health Services Research*, 15: 83 [doi: 10.1186/s12913-015-0744-0].

Garfinkel, M., Sarewitz, D. and Porter, A. (2006) A societal outcomes map for health research and policy, *American Journal of Public Health*, 96 (3): 441–6.

Halcomb, E., Davidson, P. and Hardaker, L. (2008) Using the consensus development conference method in healthcare research, *Nurse Researcher*, 16 (1): 56–71.

Inamdar, N., Kaplan, R. and Reynolds, K. (2002) Applying the balanced scorecard in healthcare provider organizations, *Journal of Healthcare Management*, 4 (3): 179–95.

Jones, J. and Hunter, D. (1995) Consensus methods for medical and health services research, *British Medical Journal*, 311 (7001): 376–80.

Mash, B., Mayers, P., Conradie, H., Orayn, A., Kuiper, M., Morais, J. et al. (2007) Challenges to creating primary care teams in a public sector health centre: a co-operative inquiry, *South African Family Practice*, 49 (1) [doi: 10.1080/20786204.2007.10873499].

Mash, B., Mayers, P., Conradie, H., Orayn, A., Kuiper, M. and Marals, J. (2008) How to manage organisational change and create practice teams: experiences of a South African primary care health centre, *Education for Health*, 21 (2): 132.

McGlynn, E.A., Kosecoff, J. and Brook, R.H. (1990) Format and conduct of consensus development conferences, *International Journal of Technology Assessment in Health Care*, 6 (3): 450–69.

Persaud, D. and Nestman, L. (2006) The utilization of systematic outcome mapping to improve performance management in health care, *Health Services Management Research*, 19 (4): 264–76.

Reason, P. (1988) The Co-Operative Inquiry Group, in P. Reason (ed.) *Human Inquiry in Action: Developments in new paradigm research* (pp. 18–39). London: Sage.

Shortell, S.M., Gillies, R.R., Anderson, D.A., Erickson, K.M. and Mitchell, J.B. (2000) *Remaking Health Care in America: The evolution of the organized delivery system*, 2nd edn. San Francisco, CA: Jossey-Bass.

Tsasis, P., Evans, J.M., Forrest, D. and Jones, R.K. (2013) Outcome mapping for health system integration, *Journal of Multidisciplinary Healthcare*, 6: 99–107.

Williams, B. and Hummelbrunner, R. (2011) *Systems Concepts in Action: A practitioner's toolkit*. Stanford, CA: Stanford University Press.

Wodchis, W.P., Dixon, A., Anderson, G. and Goodwin, N. (2015) Integrating care for older people with complex needs: key insights and lessons from a seven-country cross-case analysis, *International Journal of Integrated Care*, 15 (6) [doi: 10.5334/ijic.2249].

Suggestions for further reading

Earl, S., Carden, F. and Smutylo, T. (2001) *Outcome Mapping: Building learning and reflection into development programs*. Ottawa: International Development Research Centre.

Outcome Mapping Learning Community [http://www.outcomemapping.ca/]. Contains a resource library, examples of OM applications, discussion forums, and a community newsletter.

Shaxson, L. and Clench, B. (2011) *Outcome Mapping and Social Frameworks: Tools for designing, delivering and monitoring policy via distributed partnerships*. Delta Partnership Working Paper #1 [available from: https://www.outcomemapping.ca/resource/outcome-mapping-and-social-frameworks-tools-for-designing-delivering-and-monitoring-policies-via-distributed-partnerships; accessed 14 July 2017].

Three-part video: Sarah Earl on Outcome Mapping:
Part 1: https://www.youtube.com/watch?v=fPL_KEUawnc
Part 2: https://www.youtube.com/watch?v=a9jmD-mC2lQ
Part 3: https://www.youtube.com/watch?v=ulXcE455pj4

13 Conclusion: mapping methods to research challenges

Don de Savigny, Karl Blanchet and Taghreed Adam

This handbook brings together a suite of eleven selected methods or tool sets for health systems and policy researchers interested in applying systems thinking approaches in their research. There are, of course, many other excellent system thinking methods used both in health research and in other disciplines and these are usefully catalogued by Peters (2014) and by Williams and Hummelbrunner (2009). In this handbook, we have selected methods that collectively could be used in a spectrum across the life cycle of any research project, including deciding how the research team should be fully aware of the system they study, determining the boundaries of the system under study, identifying and analysing the stakeholders concerned, identifying and framing the system problem(s) to be studied, identifying potential solutions, addressing and negotiating decision-making processes, testing and modelling solutions, and monitoring and evaluating system-level interventions.

Figure 13.1 provides a matrix to link these research needs to the various methods chapters described in this handbook. We also indicate which methods have associated visualization features or specific software applications.

For further information on choosing methods that best fit potential research problems, see Chapter 1, where we provide a short chapter-by-chapter abstract of each approach or tool.

Taken together we hope the approaches provided in this handbook will make it easier for students, health researchers, and health system managers to move beyond systems thinking theory to real-world applications for complex health system problems. Of course, one does not need to apply all of these methods in each study, but almost every health system research endeavour would benefit from applying at least one of these systems thinking methods. Collective judgement will be needed in selecting which of the many qualitative and quantitative methods or tools described here will be most useful in particular settings. Doing so would reinforce the ability to perceive the system as a whole, and take stakeholders beyond relying on implicit mental models of how the system behaves, towards more objective and evidence-based models. Moreover, many of the tools engage stakeholders in ways that improve communicating about and collectively understanding better the complexity of the system such that negotiating, deciding, and owning interventions is made easier.

The common theme of all of these systems thinking methods is that systems behaviour is governed by common principles that can be identified and better

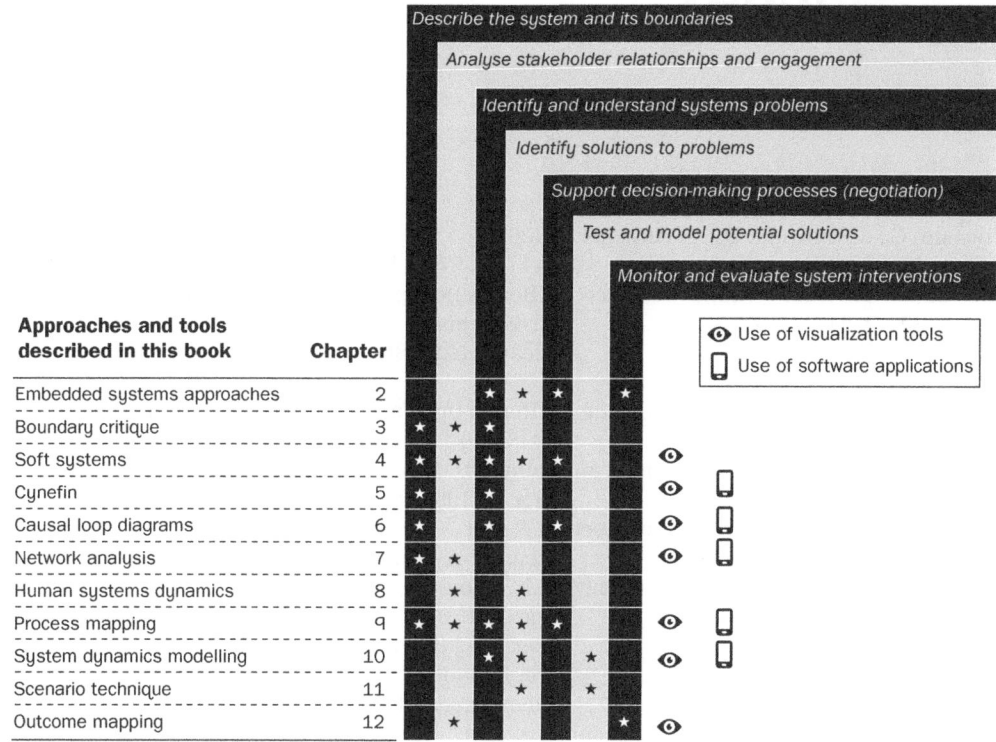

Figure 13.1 An application matrix linking research needs to respective methods chapters in this handbook

understood, and in so doing, system outcomes can be better documented and predicted, thus facilitating the development of interventions for amplification of good outcomes and the damping of bad outcomes. We hope greater application of systems thinking methodologies in health systems research will provide new opportunities to understand, communicate, test, and intervene effectively in complex health systems.

References

Peters, D.H. (2014) The application of systems thinking in health: why use systems thinking?, *Health Research Policy and Systems*, 12: 51 [doi: 10.1186/1478-4505-12-51].

Williams, B. and Hummelbrunner, R. (2009) *Systems Concepts in Action: A practitioner's toolkit.* Stanford, CA: Stanford University Press.

Index